T0330787

Urban Air Mobility
Intelligent, Safe and Sustainable Systems for Future Transportation

RIVER PUBLISHERS SERIES IN TRANSPORT TECHNOLOGY

Series Editors:

HAIM ABRAMOVICH
Technion - Israel Institute of Technology, Israel

THILO BEIN
Fraunhofer LBF, Germany

The "River Publishers Series in Transport Technology" is a series of comprehensive academic and professional books which focus on theory and applications in the various disciplines within Transport Technology, namely Automotive and Aerospace. The series will serve as a multi-disciplinary resource linking Transport Technology with society. The book series fulfils the rapidly growing worldwide interest in these areas.

Books published in the series include research monographs, edited volumes, handbooks and textbooks. The books provide professionals, researchers, educators, and advanced students in the field with an invaluable insight into the latest research and developments.

Topics covered in the series include, but are by no means restricted to the following:

- Automotive
- Aerodynamics
- Aerospace Engineering
- Aeronautics
- Multifunctional Materials
- Structural Mechanics

For a list of other books in this series, visit www.riverpublishers.com

Urban Air Mobility
Intelligent, Safe and Sustainable Systems for Future Transportation

Editors

Vishnu Kumar Kaliappan

KPR Institute of Engineering and Technology, Tamil Nadu, India

Mohana Sundaram Kuppusamy

KPR Institute of Engineering and Technology, Tamil Nadu, India

Dugki Min

Konkuk University, Seoul, South Korea

NEW YORK AND LONDON

Published 2024 by River Publishers
River Publishers
Alsbjergvej 10, 9260 Gistrup, Denmark
www.riverpublishers.com

Distributed exclusively by Routledge
605 Third Avenue, New York, NY 10017, USA
4 Park Square, Milton Park, Abingdon, Oxon OX14 4RN

Urban Air Mobility/ by Vishnu Kumar Kaliappan, Mohana Sundaram Kuppusamy, Dugki Min.

Routledge is an imprint of the Taylor & Francis Group, an informa business

ISBN 978-87-7022-678-3 (hardback)
ISBN 978-87-7004-650-3 (paperback)
ISBN 978-87-7004-643-5 (online)
ISBN 978-87-7004-634-3 (master ebook)

While every effort is made to provide dependable information, the publisher, authors, and editors cannot be held responsible for any errors or omissions.

Contents

Preface

Urban air mobility (UAM) is an emerging transportation paradigm that aims to provide fast, efficient, and sustainable mobility solutions for urban areas. This book explores the history, adoption, research, development, challenges, and mitigation strategies for UAM.

Chapter 1 focuses on the modeling and analysis of urban transportation systems, with a particular emphasis on UAM. The chapter highlights the need for accurate and efficient modeling tools to design, evaluate, and optimize UAM systems.

Chapter 2 introduces a system dynamics model of the urban transportation system, with a particular emphasis on UAM. The chapter describes the design and implementation of the model and highlights its potential applications in the field of UAM.

Chapter 3 focuses on deep learning methods for high-level control using object tracking. The chapter presents an overview of deep learning techniques for object tracking and discusses their potential applications in the field of UAM.

Chapter 4 provides a comprehensive review of deep learning models for urban aerial mobility. The chapter discusses the latest advancements in deep learning models for UAM and highlights their potential applications in the field.

Chapter 5 introduces reinforcement learning for automated electric vertical takeoff and landing decision-making for drone taxis. The chapter presents a novel approach to automated decision-making for drone taxis and discusses its potential applications in the field of UAM.

Chapter 6 discusses urban aerial mobility concepts modeling and challenges. The chapter presents a comprehensive overview of the current state of the art in UAM concepts modeling and discusses the major challenges faced by UAM systems.

Chapter 7 focuses on reinforcement learning approaches for urban air mobility/navigation and traffic control systems for UAM. The chapter presents a novel approach to reinforcement learning for UAM navigation and traffic control systems and discusses its potential applications in the field.

Chapter 8 highlights the challenges in battery charging in urban aerial mobility. The chapter presents an overview of the latest advancements in battery technology and discusses the challenges faced by UAM systems in implementing these technologies.

Chapter 9 discusses safety and security challenges in implementing urban air mobility. The chapter presents a comprehensive overview of the major safety and security challenges faced by UAM systems and discusses potential strategies for addressing these challenges.

List of Figures

List of Tables

List of Contributors

Aruna, R., *VelTech Rangarajan Dr. Sangunthala R&D Institute of Science and Technology, India*

Budiyono, Agus, *Indonesia Center for Technology Empowerment, Indonesia*

Devi, Shyamala M., *Department of Computer Science and Engineering, Panimalar Engineering College, Chennai, Tamil Nadu, India*

Dharani, J., *Department of Computer Science and Engineering, KPR Institute of Engineering and Technology, India*

Dugki, Min, *Department of Computer Science and Engineering, Konkuk University, South Korea*

Duraisamy, Premkumar, *Department of Computer Science and Engineering, KPR Institute of Engineering and Technology, India*

Goundar, Sam, *Department of Information Technology, RMIT University, Vietnam*

Jeon, Sango Woo, *Department of Computer Science and Engineering, Konkuk University, South Korea*

Kaliappan, Vishnu Kumar, *Department of Computer Science and Engineering, KPR Institute of Engineering and Technology, India*

Kanmani, P., *Department of Data Science and Business Systems, SRM Institute of Science and Technology, Kattankulathur, Chengalpattu, India*

Karthic, S., *Department of Computer Science and Engineering, KPR Institute of Engineering and Technology, India*

Kathiresan, K., *Department of Computer Science and Engineering (AIML), Sri Eshwar College of Engineering, India*

Kurmi, Sourav Patel, *School of Computer Science and Engineering, Vellore Institute of Technology, Chennai, India*

Lee, Jae-woo, *Konkuk Aerospace Design Airworthiness Institute, Konkuk University, Seoul, South Korea*

Li, Jueying, *Department of Computer Science and Engineering, Konkuk University, South Korea*

Pavithra, R., *Department of Computer Science and Engineering, NGP Institute of Engineering and Technology, India*

Prakash, P., *School of Computer Science and Engineering, Vellore Institute of Technology, Chennai, India*

Rathinasamy, Dhivya, *Department of Artificial Intelligence and Data Science, PSNA College of Engineering and Technology, India*

Sakthivel, V., *School of Computer Science and Engineering, Vellore Institute of Technology, Chennai, India*

Santhoshi, B. Kavya, *Godavari Institute of Engineering and Technology (A), India*

Sivaramakrishnan, R., *Department of Computer Science and Engineering, KPR Institute of Engineering and Technology, India*

SriPreethaa, K. R., *School of Computer Science and Engineering Vellore Institute of Technology, Vellore, India*

Sundaram, K. Mohana, *KPR Institute of Engineering and Technology, India*

Yuvaraj, N., *School of Computer Science and Engineering, Vellore Institute of Technology, Vellore, India*

List of Abbreviations

4D	Dull, dirty, dangerous, and dear
A2A	Air-to-air
A2I	Air-to-infrastructure
AAM	Advanced air mobility
ABS	Agent-based simulation
ADAS	Advanced driver assistance systems
AE	Auto-encoders
AI	Artificial intelligence
ARB	Air resources board
ATC	Air traffic control
ATFM	Air traffic flow management
ATM	Air traffic management
ATS	Air taxi services
AV	Autonomous vehicle
BPTT	Backpropagation through time
CBD	Central business district
CDR	Call detail record
CNN	Convolutional neural network
COMET	Context-aware IoU-guided tracker
CommNet	Communication neural network
CR	Close range
CSP	Cloud service provider
CV	Computer vision
DART	Dynamic autonomous rapid transit
DBN	Deep belief networks
DDoS	Distributed denial of service
DEP	Distributed electric propulsion
DL	Deep learning
DNN	Deep neural network
DoE	Design of experiments
DoS	Denial of service

DPG	Deterministic policy gradient
DQN	Deep Q-network
DRL	Deep reinforcement learning
DRONE	Dynamic remotely operated navigation equipment
DSOD	Deeply supervised object detector
DVRP	Dynamic vehicle routing problem
DVS	Dynamic vision sensor
EM	Expectation maximization
EMP	Electromagnetic pulse
EV	Electric vehicle
eVTOL	Electric vertical takeoff and landing
FHSS	Frequency hopping spread spectrum
GA	General aviation
GCS	Ground control station
GGD	Generalized graph difference
GHG	Greenhouse gas
GIS	Geographic information systems
GMM	Gaussian mixture model
GNSS	Global navigation satellite systems
GPS	Global positioning system
GPU	Graphical processing unit
HAP	High-altitude platform
HCI	human−computer interaction
HDI	Human−drone interaction
HILS	Hardware-in-the-loop
ICE	Internal combustion engine
ICEV	Internal combustion engine vehicle
IFR	Instrumental flight rules
I-MAAC	Improved MAAC
INS	Inertial navigation system
IoT	Internet of things
IPP	Integration pilot program
ITS	Intelligent transport systems
JNN	Joint neural network
LAP	Low altitude platforms
LiDAR	Light detection and ranging
LOS	Line-of-sight
LSTM	Long−short-term memory

MAAC	Multi-actor-attention-critic
MADDPG	Multi-agent deep deterministic policy gradient
MADRL	Multi-agent deep reinforcement learning
MAP	Medium altitude platforms
MAS	Multi-agent systems
MDP	Markov decision process
ML	Machine learning
MLP	Multilayer perceptron
MOT	Multiple objects tracking
MPC	Model predictive controller
MR	Medium range
MRE	Medium range endurance
MRP	Markov reward process
MTOW	Maximum takeoff weight
NLOS	Non-line-of-sight
NOMA	Non-orthogonal multiple access
NOVPC	Number of vehicles per capita
OD	Origin−destination
ODM	On-demand mobility
OWT	Offshore wind turbine
PAM	Personal air mobility
PAV	Personal air vehicle
PT	Public transportation
RAM	Rural air mobility
Re-ID	Re-identification
RF	Radio frequency
RGB	Red, green, and blue
RL	Reinforcement learning
RMSE	Root mean square error
RNN	Recurrent neural network
RPA	Remotely piloted aircraft
RPAS	Remotely piloted aerial system
RTT	Research transition team
SAR	Search and rescue
SD	System dynamics
SEL	Sound exposure level
SfM	Structure from motion
SILS	Software-in-the-Loop
SNR	Signal-to-noise ratio

SoI	Systems of interest
SoS	System of systems
SOT	Single object tracking
SR	Short range
sUAS	Small unmanned aircraft system
TBD	Tracking by detection
TNC	Transportation network company
UAM	Urban air mobility
UAS	Unmanned aircraft system
UATS	Urban air taxi services
UAV	Unmanned aerial vehicle
UML	UAM maturity level
US	Unmanned systems
UTM	Unmanned aircraft system traffic management
V2V	Vehicle-to-vehicle
VFR	Visual flight rules
VKT	Vehicle kilometer traveled
VLL	Very low level
VOT	Visual object tracking
VTOL	Vertical takeoff and landing
ZEV	Zero emissions vehicle

1

Toward Future Transportation: History, Adoption, Research, and Development, Challenges in Urban Aerial Mobility

R. Sivaramakrishnan[1], Dhivya Rathinasamy[2], Agus Budiyono[3], and Min Dugki[4]

[1]Department of Computer Science and Engineering, KPR Institute of Engineering and Technology, India
[2]Department of Artificial Intelligence and Data Science, PSNA College of Engineering and Technology, India
[3]Indonesia Center for Technology Empowerment, Indonesia
[4]College of Computer Science and Information Systems, Pacific States University, USA
E-mail: sivaraamakrishnan2010@gmail.co; dhivya.rathinasamy@gmail.com; budiyono@alum.mit.edu; dkmin@konkuk.ac.kr

Abstract

With increasing urbanization, the demand for efficient and sustainable transportation solutions has become critical. This chapter examines the different modes of transportation that exist today and the challenges facing the transportation industry, including the impact of COVID-19 on public transportation. The concepts driving the future of transportation, such as smart mobility, electric vehicles, and connected and autonomous vehicles, are discussed. Deep learning in transportation systems is also introduced as a promising approach to improving transportation safety and efficiency. Furthermore, the chapter emphasizes the importance of smart cities in future transportation and presents a comprehensive review of UAM, a rapidly emerging concept of future transportation that aims to address increasing congestion and transportation challenges in urban areas. The relevance of UAM in smart cities and current research and development efforts were highlighted. The challenges faced in the adoption of UAM were examined, which are valuable insights for policymakers and practitioners seeking to gain a better understanding of this emerging transportation system. The importance of continued research and direction is emphasized to ensure the intelligent, safe, and sustainable implementation of UAM in the future.

Keywords: Urban air mobility, future transportation, smart cities, autonomous vehicles, air traffic management, sustainable transportation

1.1 Introduction to Future Transportation

From ancient times, humans have been searching for means to enhance transportation by making it more efficient and accessible. Initially, the wheel was invented and, subsequently, carts and wagons, steam power, and the internal combustion engine were developed. Advancements have continued with the creation of electric cars, bikes, and autonomous vehicles. The implementation of concepts that were once considered to be confined to the realm of science fiction is now being actualized [1].

The evolution of transportation has been restricted in both size and range throughout human history. However, over the past two centuries, the introduction of mechanized transportation has led to significant enhancements in the capacity, speed, efficiency, and geographical coverage of transport systems. The fundamental objective to improve transport technology is to facilitate the swift, safe, and efficient movement of passengers and goods, in greater quantities. Transportation modifications can be either gradual or revolutionary. The future of transportation will be impacted by the increased integration of physical and information systems. Remarkable advancements have occurred in modes, terminals, and networks, which can be classified into two functional aspects [2]:

- Revolutionary changes: It refers to the introduction of entirely new technology that brings about fresh opportunities for transportation and the economy, often rendering existing modes of transportation obsolete due to the significant cost, capacity, or time benefits of the new mode. These types of changes are rare but have a profound impact since they usually require the establishment of entirely new networks. Although it is difficult to predict revolutionary changes, their potential impact can be evaluated once they occur. However, during the initial phase of their introduction, the potential benefits of these innovations may be overstated, resulting in excessive investments in technologies that have limited market potential and profitability.

- Incremental changes: Incremental changes, also known as evolutionary changes, refer to the gradual improvement of existing transportation technology and operations. This results in increased productivity, greater capacity, reduced costs, and enhanced performance of the involved mode or terminal. Such changes can be attributed to infrastructure and vehicle enhancements or the application of information technologies to manage operations more efficiently. While it is possible to predict incremental changes, it is challenging to determine the rate of change they bring.

1.2 Modes of Transportation

Modes of transportation are crucial constituents of transportation systems as they facilitate mobility. Modes can be categorized into three broad groups based on the medium they utilize: land, water, and air. Each mode has unique characteristics and prerequisites, and is tailored to meet the specific demands of passenger and freight traffic. As a result, significant variations in the deployment and usage of transportation modes have emerged in various parts of the world. In recent times, there has been a growing inclination toward the integration of modes through intermodality, and the linkage of modes with production and distribution activities. Nonetheless, there is also a noticeable trend toward the segregation of passenger and freight activities across most modes [3].

1.2.1 Road transportation

Road transportation encompasses a variety of motorized and non-motorized options for mobility, mainly for short distances, which users may select based on factors such as affordability, convenience, availability, and comfort. Although automobiles have become a preferred mode of passenger transportation due to their flexibility and convenience, they also contribute to traffic congestion, particularly in urban areas. However, sustainable transportation systems emphasize the significance of walking, cycling, and emerging forms of personal mobility, such as electric scooters, as crucial elements of short-distance mobility.

1.2.2 Rail transportation and pipelines

Railways consist of a designated path on which wheeled vehicles operate. With recent technological advancements, rail transportation now includes monorails and maglev as well. Rail systems generally have a moderate level of physical limitations and require a low gradient, particularly for freight. Historically, heavy industries have been closely associated with rail transport systems, but containerization has enhanced the flexibility of rail transportation by connecting it with road and maritime modes. Rail is the land transportation mode that offers the highest capacity, with a fully loaded coal unit train weighing 23,000 tons being the heaviest load ever transported. However, the gauges of rail systems vary worldwide, which can pose challenges for integration.

1.2.3 Maritime transportation

Maritime transportation is an efficient mode of transportation for moving large amounts of cargo over long distances due to its physical characteristics such as buoyancy and limited friction. The main maritime routes consist of oceans, coasts, seas, lakes, rivers, and channels. However, due to the location of economic activities, maritime transportation occurs in specific parts of the maritime space, mainly over the North Atlantic and the North Pacific. Efforts to facilitate maritime circulation include constructing channels, locks, and dredging to reduce discontinuity. However, such endeavors are expensive. Comprehensive inland waterway systems include Western Europe, the Volga/Don system, the St. Lawrence/Great Lakes system, the Mississippi and its tributaries, the Amazon, the Panama/Paraguay, and the interior of China.

1.2.4 Air transport

Air transportation is highly flexible as it covers extensive areas, but its density is highest over the North Atlantic, inside North America and Europe, and over the North Pacific. Air transport faces various constraints, such as the airport site requiring a minimum runway length of 3300 meters for takeoff and landing, weather conditions, fog, and air currents. The tertiary and quaternary sectors, such as finance and tourism, heavily rely on air travel for long-distance mobility. Additionally, air transport has been playing a growing role in global logistics, with increasing amounts of high-value freight being transported through air cargo.

1.2.5 Intermodal transportation and containerization

Intermodalism refers to the use of multiple transportation modes together to maximize the benefits of each mode. While this concept can be applied to passenger transportation, it has primarily impacted freight transportation.

Figure 1.1 Different modes of transportation.

By utilizing containerization, which allows for standardized shipping containers to be easily transferred between maritime and land transportation systems, intermodal integration has become increasingly feasible. Figure 1.1 exemplifies the different modes of transportation.

1.3 Challenges for Transportation

The transportation sector is gaining more importance and undergoing changes due to various challenges and factors, including sustainability, congestion, governance, and technology. With the increasing complexity of the transport industry, it is necessary to replace conventional approaches that focus on a limited range of factors with more nuanced analysis and solutions. Freight mobility is becoming more critical in the discipline due to the growth of urban freight distribution and global supply chains. The scope of transport geography remains diverse in the transport industry, public planning, and research institutions. Future transportation systems will likely be shaped by similar forces as in the past, but it is uncertain which technologies will dominate and how they will impact the spatial structure [4].

1. Improving transport infrastructure: The transportation sector is facing the challenge of outdated and insufficient infrastructure, which often leads to congestion, delays, and safety issues. To tackle these challenges, there is a need for significant investment in modernizing and expanding transportation infrastructure, including roads, bridges, railways, and airports, to ensure reliable and efficient transportation services for people and goods.

2. Governance and management: Effective governance and management are crucial for the transportation sector to operate efficiently and meet

the needs of its users. However, the industry is facing challenges related to regulatory frameworks, funding, and stakeholder coordination. To overcome these challenges, there is a need for a comprehensive approach that involves all stakeholders, including government agencies, private companies, and communities, to work together toward common goals and objectives.

3. Social and environmental responsibility: The transportation sector has a significant impact on society and the environment, including air and noise pollution, greenhouse gas emissions, and social equity issues. To address these challenges, there is a need for a shift toward sustainable and socially responsible transportation systems that promote accessibility, affordability, and environmental stewardship. This requires a balance between economic, social, and environmental objectives, and the involvement of all stakeholders to ensure their needs are met.

4. Future transportation systems: The transportation sector is experiencing rapid technological advancements that are leading to the development of new transportation systems, such as autonomous vehicles, electric cars, and urban aerial mobility. These new systems offer the potential to revolutionize the way people and goods move around, but they also present new challenges related to infrastructure, regulation, and social acceptance. To successfully implement future transportation systems, there is a need for innovative solutions that address these challenges while ensuring the safety, reliability, and sustainability of transportation services.

1.4 COVID-19 and Public Transportation

The ongoing COVID-19 pandemic is causing widespread disruptions in various aspects of life, including transportation. The effects of the pandemic on people's mobility and urban freight movements highlight the inadequacy and limitations of traditional transportation systems in dealing with dynamic urban traffic. To address these challenges, transportation operators should view these disruptions as opportunities to reassess the issues that have been exposed by the pandemic and take proactive measures to improve the transportation system. One example of limitations exposed by the pandemic is the lack of dynamic organization in the transportation system, heavy reliance on workforce staffing, and inability to convert passenger vehicles for freight transport. These shortcomings highlight the need to reflect on how to improve the current transportation systems to cope better with urban traffic in the future. In addition, potential changes in travel patterns after the pandemic should be considered to enhance the current transportation services and inform the design of future transportation systems.

As the world advances technologically, urban transportation systems are facing both benefits and drawbacks. By making use of emerging technologies, such as information and communications technology (ICT) and autonomous vehicle (AV), urban transportation challenges can be better addressed and resolved for the future. This study compiles data on how the COVID-19 pandemic has affected urban transportation of both people and goods. To address long-term challenges, an integrated system called dynamic autonomous rapid transit (DART) is proposed as a means to transport both people and freight. The technical obstacles of designing and implementing such a system, as well as the corresponding business model, are thoroughly examined [5].

The integration of the DART system is a top priority for urban transportation. By granting priority to DART freight modules in public transport operations, the benefits of integrating public transport and urban freight delivery can be realized.

1.5 The Concepts Driving the Future of Transportation

The development of transportation in the future requires the adoption of innovative and intelligent energy sources, transportation modes, and physical and technological infrastructures that will support and enable these transportation advancements [1, 6].

The common themes in transportation innovation are:

1.5.1 Smart technology

The application of smart transportation and smart city traffic management is transforming how cities address mobility and emergency response while mitigating congestion on city streets. This is achieved through the use of sensors, advanced communication technologies, automation, and high-speed networks.

Transportation, the process of moving from one location to another, has been a fundamental aspect of human existence throughout history. From ancient times when chariots and horses were used to the modern era where carriages, automobiles, steam trains, and spacecraft have been developed, mobility is a significant part of the human experience.

1.5.2 Electrification

As the demand for electric vehicles (EVs) grows, there is an increasing trend in the availability of all-electric trucks, buses, and industrial/agricultural vehicles, in addition to personal vehicles. Large corporations such as Amazon are taking steps toward an all-electric fleet by investing in thousands of electric delivery trucks, while the US Postal Service is also planning to introduce a fleet of EVs.

The growth of electric vehicles is not just limited to personal cars, but also expanding to include trucks, buses, and industrial/agricultural vehicles. Major companies like Amazon and the US Postal Service have plans to purchase thousands of electric delivery trucks, with the goal of transitioning to an all-electric fleet in the near future. Furthermore, the trend of electrification is extending to various fields such as agriculture equipment and scooters. Electric buses and trains have been in use for decades, but we are now witnessing the emergence of electric long-haul trucks that could transform the landscape of interstate commerce. Moreover, advancements in powerful and lightweight electric motors are enabling the electrification of aircraft as well.

1.5.3 Autonomy

Autonomous transport systems refer to the self-driving transfer of equipment, people, or resources from one point to another with minimal human intervention. These systems encompass various types of transport vehicles such as buses, trucks, trains, metros, ships, and airplanes. Currently, they are mostly utilized in controlled industrial areas but are anticipated to be used in public areas soon, with varying levels of autonomy.

1.5.4 Solar panel roads generating electricity

Solar technologies are advancing rapidly with numerous innovative concepts and ideas being introduced to the market. Some countries have been experimenting with using solar panels as the surface material for roads.

The implementation of solar road paving has been initiated in some areas of the Netherlands, and the country aims to gather practical knowledge regarding the maintenance and management of such roads. The projected output of the solar road is around 30,000 kWh per year for every 100 m of road.

Solar panel roads, like those in the Netherlands, can improve renewable energy and sustainable urban infrastructure by generating electricity, powering streetlights and other infrastructure, charging EVs, and melting ice and snow.

1.5.5 AI and data-driven maritime transport

The development of AI is a significant factor shaping the future of transportation in all its forms. Norway and Finland have demonstrated the feasibility of autonomous ships, which are anticipated to become available in the market soon.

Advanced AI has the potential to compute optimized routes and transport speed by combining weather and sea current information. The vast amount of data generated by ships through their smart systems could represent a significant breakthrough in the shipping industry since the introduction of containers.

1.5.6 Hydrogen economy for greener transport

Hydrogen is considered an environmentally friendly fuel due to its abundant availability and lack of pollution, but there are challenges in using it as an energy source.

These challenges include the need for manufacturing as it does not occur naturally in a usable form, and the manufacturing process requires significant amounts of electricity, mostly produced by fossil fuels. Furthermore, storing and transporting hydrogen can be hazardous because of its explosive nature.

1.5.7 Road traffic as software

With the advent of autonomous vehicles, a centralized software system is expected to emerge that will monitor and control the entire traffic system. The rationale behind this is the potential conflict that may arise between the different AI algorithms used by various manufacturers.

Such software would prevent crashes and congestion, as well as improve safety. By forming an adaptive system, which considers the needs of both passengers and goods, AI can play a crucial role in optimizing traffic management.

1.5.8 High-speed travel

Various transportation systems, such as bullet trains and transport capsules, are under development to provide unparalleled travel speeds. One such example is Hyperloop, which is gaining popularity and being considered for potential routes globally.

1.5.9 Lightweight vehicles

Vehicle manufacturers are striving to produce high-performance and efficient vehicles. Research indicates that a 10% reduction in vehicle weight can improve fuel efficiency by 6% or more. To achieve this, manufacturers are exploring the use of alternative materials such as carbon fiber and magnesium-aluminum alloys to replace heavier metals like iron and steel. However, more work is needed to realize this concept [7].

1.5.10 Delivery drones

Various companies including Amazon, UPS, and DHL are investing in the development of delivery drones, which is proving to be a significant technological challenge. These companies are committed to advancing drone technology in order to make it a practical solution for fast mile delivery [7].

1.5.11 Bicycle sharing system

Bicycle sharing systems can provide an alternative for those who want to avoid main transport hubs. Users can rent a bicycle from the start of their trip and return it at their destination. Electric bicycles can be used for various deliveries by businesses, which can reduce costs. Studies indicate that bicycle sharing can significantly reduce traffic and pollution [7].

Given the significant growth of these technologies in recent times, it is reasonable to expect that they will all play crucial roles in shaping the future of transportation.

1.6 Deep Learning in Transportation Systems

The rise of machine learning has replaced various statistical models and provided better problem-solving capabilities in multiple fields of study. The growth of machine learning has also impacted transportation systems, especially intelligent transportation systems (ITS). The increasing availability of data and advancements in computational techniques like graphical processing units (GPUs) have made deep learning (DL), a subset of machine learning, popular in this domain.

Deep learning (DL) has become a promising solution in the field of intelligent transportation systems (ITS) due to its ability to handle vast amounts of data and extract knowledge from complex systems. DL has a range of network types that allow researchers to approach problems with neural network techniques. This has led to new solutions for various transportation engineering problems such as traffic signal control, transportation security through surveillance sensors, traffic rerouting, health monitoring of transportation infrastructure, and other challenging problems. DL has become a powerful and effective method to tackle these issues in the transportation industry. The following are the various applications of DL in ITS [8].

1.6.1 Traffic characteristics prediction

In the field of transportation, predicting traffic characteristics using DL is an area of great interest. Predicting traffic flow, speed, and travel time can help drivers make informed decisions about their routes, and traffic management agencies can use the information to manage traffic more effectively. As these characteristics are interrelated, methods used to predict one can be applied to predict the others as well.

1.6.2 Traffic incident inference

In the realm of transportation, predicting the risk of traffic incidents in a particular location and detecting incidents based on traffic features are critical objectives for traffic management agencies. These goals aim to minimize incident risk in hazardous areas and prevent traffic congestion in incident locations. Although factors such as driver behavior may be unpredictable, certain key features can aid in forecasting traffic incidents.

1.6.3 Vehicle identification

Re-identification (Re-ID) has a broad range of applications, from estimating travel time to automated ticketing. The initial step in Re-ID is license plate recognition, given that license plates are unique to each vehicle.

1.6.4 Traffic signal timing

Intelligent transportation system (ITS) management involves multiple types of data to control traffic flow, and one of its key tasks is optimizing traffic signal lights. In the transportation field, finding the optimal timing for signal lights has been a challenge for many years. Researchers have developed analytical models that use mathematical techniques to solve this optimization problem. However, with the emergence of deep learning (DL), the approach to modeling traffic dynamics has taken a new direction. RL, in particular, has proven useful in finding the best traffic signal timing, and its application in various studies has become more feasible due to its nature.

1.6.5 Visual recognition tasks

The application of DL has been highly effective in the development of non-intrusive systems, such as camera-based recognition and detection systems. These systems have diverse applications, ranging from providing suitable roadway infrastructure for driving vehicles to enabling autonomous vehicles to have a safe and reliable driving strategy.

1.6.6 Ride sharing and public transportation

Public transportation is a crucial aspect of urban mobility, with bus and metro systems, taxis, and other modes of transport facilitating passenger movement. Enhancing city planning efficiency and passenger satisfaction have been a major focus for transportation companies, with DNN providing increasingly optimal routing maps. These maps consider passenger demand for a particular mode of travel at specific locations and times. By adopting DL techniques, companies can achieve even greater accuracy in predictions compared to existing ML methods.

1.7 The Future Transportation in Smart Cities

To enhance the planning, construction, management, and services of cities, smartcities utilize modern information technologies, such as cloud computing, big data, space geographical information integration, and the Internet of Things. Transportation is a crucial component of smart cities, and it involves smart roads, smart street lights, smart traffic signs, and smart cars [8]. Here are some possible transportation options for future smart cities.

1.7.1 Electric and autonomous vehicles

These vehicles are expected to dominate the streets of smart cities, providing safe and efficient transportation. They will be equipped with advanced technologies, such as sensors and communication systems, to navigate roads and avoid accidents.

1.7.2 Public transportation

Public transportation systems will be more advanced, incorporating new technologies such as smart payment systems, real-time route optimization, and automated maintenance.

1.7.3 Electric and autonomous vehicles

These vehicles are expected to dominate the streets of smart cities, providing safe and efficient transportation. They will be equipped with advanced technologies, such as sensors and communication systems, to navigate roads and avoid accidents.

1.7.4 Public transportation

Public transportation systems will be more advanced, incorporating new technologies such as smart payment systems, real-time route optimization, and automated maintenance.

1.7.5 Bike-sharing and scooter-sharing

Shared bikes and scooters are becoming increasingly popular in urban areas. In smart cities, these systems will be more efficient and connected to the transportation network, making it easier for people to get around.

1.7.6 Drones and aerial vehicles

Drones are now being extensively researched and utilized due to their versatility and ability to be applied in numerous fields, including security, monitoring, and exploration of hard-to-reach areas [9]. As technology advances, drones and aerial vehicles may become a more common form of transportation, particularly for goods delivery and emergency services.

1.7.7 Hyperloop and high-speed trains

Hyperloop technology and high-speed trains are being developed to provide faster and more efficient transportation between cities.

1.7.8 Pedestrian and bike-friendly infrastructure

Smart cities will prioritize creating pedestrian and bike-friendly infrastructure to encourage more active modes of transportation and reduce traffic congestion. This includes wider sidewalks, bike lanes, and green spaces.

1.8 Introduction to Urban Aerial Mobility (UAM)

1.8.1 UAM: An introduction

Urban air mobility (UAM) is an emerging concept that envisions air transportation as an integrated part of the urban transportation system. It refers to the use of electric vertical takeoff and landing (eVTOL) aircraft and other aerial vehicles to transport passengers and cargo within urban and suburban areas, providing faster and more efficient transportation options for urban dwellers. UAM is seen as a potential solution to the increasing traffic congestion on the ground and the demand for more sustainable and efficient transportation.

1.8.2 UAM history

UAM has evolved over the years with different phases marked by changes in technology and market demand. The history of UAM can be traced through six distinctive phases [10], starting from the conceptual phase of flying cars to the current phase of point-to-point air taxi services. Each phase has brought about new innovations, new players, and new opportunities in the UAM industry.

- Phase 1: Flying car concepts − This phase began in the early 1900s and was characterized by various attempts to create a flying car, with inventors and engineers creating various designs and prototypes. It was largely unsuccessful due to technological limitations.
- Phase 2: Early UAM operations with scheduled helicopter services − This phase started in the 1950s and lasted until the 1970s. It was marked by the introduction of scheduled helicopter services for urban commuting.
- Phase 3: Re-emergence of on-demand services − This phase began in the 2000s, marked by the emergence of on-demand air taxi services using helicopters and small fixed-wing aircraft.
- Corridor services using VTOL − In this phase, which started in the 2010s, vertical takeoff and landing (VTOL) aircraft are used to provide on-demand services within designated corridors, connecting suburbs to urban centers.
- Hub and spoke services − This phase is characterized by the use of larger UAM vehicles to connect suburban areas to transportation hubs, such as airports or train stations.

- Point-to-point air taxi services − This phase involves the use of small, on-demand aircraft to provide point-to-point transportation within urban areas. This phase is currently in development and is expected to become more prevalent in the coming years.

1.8.3 Current research and developments in UAM

The UTM initiative, which aims to establish a system for managing unmanned aircraft systems (UAS) in low-altitude airspace, is being coordinated by a research transition team (RTT) consisting of the FAA, NASA, and industry partners. The RTT is focused on developing concepts and use cases, designing information architecture, improving communication and navigation systems, and enhancing sense-and-avoid capabilities. Through research and testing, the team aims to identify the necessary requirements to ensure safe UAS operations in low-altitude airspace, both within and beyond the pilot's line of sight [11].

1.8.3.1 Design and technological requirement

To define the urban airspace size, capacity, and geometry, it is necessary to identify the structural factors that affect it. These factors can be divided into four categories: safety-related factors, social factors, system factors, and aircraft factors.

1.8.3.2 UAM concept and classification

The UAM concept encompasses regulations, processes, and technologies that allow air traffic operations for transporting goods and people in urban areas. ICAO divides airspace into two categories, controlled and uncontrolled airspace, using seven classes (A, B, C, D, E, F, and G), based on flight requirements and air traffic services provided. Classes A−E are controlled airspace, while classes F and G are uncontrolled airspace. Each class of airspace has specific guidelines that detail how aircraft should operate and how ATC should interact with them [12].

1.8.3.3 UAM traffic management

In the 1920s, the first air traffic control rules were established. Today, the air traffic management (ATM) system consists of air traffic control (ATC), air space management (ASM), air traffic flow management (ATFM), and air traffic services (ATS). As the number of flights increased, the ATM system expanded to include air route traffic control centers and airport traffic control towers. Fixed-wing aircraft have been the dominant users of airspace in ATM since its early days. However, integrating helicopters and other vertical takeoff and landing vehicles (VTOL) into air traffic flow is a challenge due to their unique performance characteristics, which differ from those of fixed-wing aircraft, resulting in suboptimal use of airport capacity [13].

1.8.3.4 Integration to existing transport system

The main challenges and research and innovation needs for the complete integration of drones are listed below [14]:

- Maturing, validating, and deploying the basic U-space services (U1 and U2) and developing advanced U-space services (U3 and U4) to enable UAS/UAM missions in highly congested and complex scenarios.
- Developing concepts and solutions for the integration of autonomous operations in densely populated and complex airspace environments to enable UAM.
- Defining systems and interfaces for seamless integration of ATM, UAM, and U-space.
- Developing concepts and solutions considering social acceptance, environmental impacts, and sustainability (such as UAM noise, visual pollution, privacy, emissions, and recycling/resource management).
- Elaborating concepts for U-space application above the very low level (VLL) airspace.

1.8.3.5 Potential users and public acceptance

- EU citizens have a positive attitude toward UAM and see it as an attractive means of mobility.
- Use cases that benefit the community, such as medical or emergency transport, are better supported than those satisfying individual/private needs.
- The main benefits expected from UAM are faster, cleaner, and extended connectivity.
- Citizens want to limit their own exposure to risks related to safety, noise, security, and environmental impact when reflecting on potential UAM operations in their city.
- Safety concerns are the top priority, but citizens trust current aviation safety levels and would be reassured if these levels were applied for UAM.

- Noise is the second main concern expressed, with the level of annoyance varying with the familiarity of the sound, distance, duration, and repetition.
- UAM is seen as a good option to improve the local environmental footprint, but citizens express major concerns about its impact on wildlife.
- There is limited trust in the security and cybersecurity of UAM, requiring threat-prevention measures.
- The integration of UAM into the existing air and ground infrastructure must respect residents' quality of life and the cultural heritage of old European cities.
- Local residents and authorities want to engage and play an active role in the implementation of UAM.

1.9 Adoption of UAM

UAM adoption is being driven by the need to improve urban mobility and address congestion, but is hindered by challenges related to infrastructure, regulations, public perception, and technology maturity. However, efforts are being made by industry players, governments, and regulatory bodies to address these challenges and accelerate the adoption of UAM [15].

1.9.1 Factors influencing the perception of aerial vehicles

NASA conducted recent studies to investigate the potential market for UAM and its implementation barriers. The studies focused on various UAM use cases in US cities, including last-mile delivery, air metro, and air taxi, and identified safety, privacy, environmental impact, noise and visual pollution, cybersecurity, affordability, and willingness to pay as major concerns. Fu et al. explored transportation mode preferences in an UAM environment and found that travel time, cost, and safety significantly impacted UAM adoption. MacSween's research on unmanned aerial vehicles found emotional and safety data to increase users' persuasions of UAV applications, including commercial, cargo, and passenger transportation. Peeta et al. predicted the probability of individuals switching to on-demand air service based on travel distance, service fare, and level of accessibility. Cost was found to be crucial in determining consumer adoption in Germany, with inter-city applications or thin-haul flights highlighting cost and time as significant factors for consumer adoption of on-demand aerial mobility. Overall, societal and environmental impacts, cost, and time were found to be key determinants of UAM adoption.

1.9.2 Factors influencing acceptance of automation in transportation

The research in automation has explored social acceptance in various contexts, including ground autonomous vehicles. Factors that influence the acceptance and trust of autonomous vehicles include reliability, safety, locus of control, perceived benefits, ease of use, and control. Real-life tests, increased transparency, and manufacturers' reputation can also help build trust in autonomous vehicles. Perceived benefits of automation can lead to higher perceived usefulness and acceptance. The attributes of time, costs, comfort, cleanliness, weather conditions, social behavior, and driving enjoyment also influence user acceptance. Social attitudes and cultural values play a role in the adoption of autonomous vehicles, and socio-demographic factors such as age, gender, and education level also impact automation perception.

1.10 Summary

UAM is a rapidly evolving concept in the transportation industry that aims to tackle the rising issues of congestion and transportation challenges in urban areas. This chapter has provided a comprehensive overview of the current state and future prospects of transportation in urban areas. The challenges facing the industry have been outlined, including the impact of the COVID-19 pandemic on public transportation. Emerging concepts such as smart mobility, electric vehicles, and connected and autonomous vehicles have been discussed, along with the potential of deep learning techniques to improve transportation safety and efficiency. The chapter has also highlighted the importance of smart cities in the future of transportation, and presented a thorough examination of UAM as a promising solution to address the growing congestion and transportation challenges in urban areas. The challenges of adopting UAM have been discussed, and potential mitigation strategies have been presented, providing valuable insights for policymakers and practitioners. It is clear that continued research and development will be essential to ensure the successful implementation of UAM and other emerging transportation solutions.

References

[1] Conrad Galambos, 'The future of transportation: Where will we go?', 2019, https://www.geotab.com/blog/future-of-transportation/

[2] Dr. Jean Paul Rodrigue, 'Future Transportation Systems', https://transportgeography.org/contents/conclusion/future-transportation-systems/

[3] Main Passenger Modal Options, https://transportgeography.org/contents/chapter5/transportation-modes-modal-competition-modal-shift/passenger-modal-options/

[4] Challenges for Transport Geography, https://transportgeography.org/contents/conclusion/

[5] Sun, S., Wong, Y. D., Liu, X., & Rau, A., 'Exploration of an integrated automated public transportation system', Transportation Research Interdisciplinary Perspectives, 8, 100275, 2020.

[6] Shiori Ota, Marianna Mäki-Teeri, Max Stucki, '12 Trends That Will Drive the Future of Transport', 2020, https://www.futuresplatform.com/blog/12-trends-will-drive-future-transport

[7] Mayur Panchal, '6 Future Transportation Technologies To Look Forward In Future', 2019, https://yourstory.com/mystory/six-future-transportation-technologies-future/amp

[8] Haghighat, A. K., Ravichandra-Mouli, V., Chakraborty, P., Esfandiari, Y., Arabi, S., & Sharma, A., 'Applications of deep learning in intelligent transportation systems', Journal of Big Data Analytics in Transportation, pp. 115-145, 2(2), 2020.

[9] Saeed H. Alsamhi, Ou Ma, M. Samar Ansari and Sachin Kumar Gupta, "Collaboration of Drone and Internet of Public Safety Things in Smart Cities: An Overview of QoS and Network Performance Optimization", Drones, MDPI, 2019.

[10] Cohen, A. P., Shaheen, S. A., & Farrar, E. M. (2021). Urban air mobility: History, ecosystem, market potential, and challenges. IEEE Transactions on Intelligent Transportation Systems, 22(9), 6074-6087.

[11] Edwin Vattapparamban, Ismail Guvenc, Ali I, Yurekli, Kemal Akkaya, and SelcukUluagacÂÿ Department of Electrical and Computer Engineering, Florida International University, Miami, FL, USA, "Drones for Smart Cities: Issues in Cybersecurity, Privacy, and Public Safety", International Wireless Communications and Mobile Computing Conference (IWCMC), 2016.

[12] KhaulaAlkaabi and Abdel Rhman El Fawair, "Drones applications for smart cities: Monitoring palm trees and street lights", GE GRUYTER, 2022.

[13] Aleksandar Bauranov a, JasenkaRakas , "Designing airspace for urban air mobility: A review of concepts and approaches", Progress in Aerospace Sciences, Volume 125, 1 August 2021.

[14] Bianca I. Schuchardt, DagiGeister, Thomas Lüken, Franz Knabe, Isabel C. Metz, NiklasPeinecke and Karolin Schweiger, "Air Traffic Management as a Vital Part of Urban Air Mobility—A Review of DLR's Research Work from 1995 to 2022", Aerospace, MDPI, 2022.

[15] Al Haddad, C., Chaniotakis, E., Straubinger, A., Plötner, K., & Antoniou, C. (2020). Factors affecting the adoption and use of urban air mobility. Transportation research part A: policy and practice, 132, 696-712.

2

Modeling and Analysis of Urban Transportation

V. Sakthivel[1], Sourav Patel Kurmi[1], P. Prakash[1], and Jae-Woo Lee[2]

[1]School of Computer Science and Engineering, Vellore Institute
of Technology, Chennai, India
[2]Konkuk Aerospace Design Airworthiness Institute, Konkuk University,
Seoul, South Korea
E-mail: mvsakthi@gmail.com; souravkurmi5683@gmail.com;
prakash.p@vit.ac.in; jwlee@konkuk.ac.kr

Abstract

Road traffic is one of the biggest problems in our modern society. In Europe, for example, 4 in 10 citizens experience daily, lengthy commuting problems. Together with high fuel consumption, noise problems, and carbon emissions, these will create high costs. Do you not think it is time to change this situation? It could be that we have a solution to all these problems. Urban air mobility is a vision taking shape now. It seems determined to transform our cities into cities of the future. Maintaining a good traveling speed for cars makes the urban environment particularly dangerous, especially when road traffic interacts with cyclists and pedestrians. Safety remains a challenge in the urban center. This chapter brings good insight into the modeling and analysis of urban transportation systems. This chapter discusses the concept of airspace, its strengths and weaknesses, and a well-defined approach to designing urban airspace. What is the requirement for urban aviation and how is the UAM market divided into sub-markets and market actors? Finally, this chapter discusses a few UAM modeling approaches. Intelligent transportation system (ITS) modeling deals with two principles: the management of travel and transportation demand and supply. As a future perspective of a sustainable environment, one must need a new perspective toward urban air mobility and make enhancements including the perspective of urban planning.

Keywords: Urban air mobility, urban transportation, personal air mobility, personal air vehicle

2.1 Introduction: Modeling of UAM

UAM is a term used to describe a system of urban air transportation [1]. It refers to conventional and still-in-use machinery including helicopters, VTOLs, and various unmanned aerial vehicles. These aircraft propel themselves and maintain velocity using several electric-driven blades [2]. The main objective of urban air mobility, in contrast to the surface and underground transportation systems that currently control the majority of urban mobility, is to use airspace as the third dimension by flying vehicles, according to the website of the European Union project on smart cities. Similar to urban air mobility (UAM), the terms personal air mobility (PAM), personal air vehicle (PAV), and advanced air mobility (AAM) are widely used.

Imagine not having valuable assistance like Google Maps. You would easily lose orientation; it would require a global view of the city, which is possible only from above it. This approach was imagined more than 2000 years ago. So the concept of freedom through a third dimension was already envisioned at that time. Why is the topic of urban air mobility growing so rapidly?

Let us have a look at some population breakdowns and statistical information together: The percentage of the world's population that dwells in urban regions is consistently going up, and this trend is expected to continue. In 2007, the percentage of old persons represented more than three-quarters of the total population in countries with lower levels of economic development. It is anticipated that this percentage would reach 80% by the year 2013. Even though urbanization is proceeding at an even more rapid pace in Africa, just a small percentage of the continent's population lives in urban areas. One example would be the continent of Africa. There has been a significant rise in mobility as a direct result of the quick rate at which urbanization is occurring. Problems of a significant kind may be found in metropolitan regions. One of the most well-recognized problems in Metropolis is the city's inadequate capacity for the flow of motor traffic. Because the daily demand for mobility is more than the capacity of the roads and subways, the consequence is that the transport system gets saturated during peak hours, resulting in a decline in mobility. This is the effect of the saturation of the transport system. High vehicle speeds in urban areas create a hazardous environment, particularly when road traffic interacts with cyclists and pedestrians. Safety remains a challenge in the urban center, surface mobility, and internal combustion. Engines are responsible for aggravating air pollution in large cities; the limits suggested by the Public Health Organization are frequently exceeded. This is a clear signal that sustainable development of urbanization cannot be based on the massive use of cars.

The development of these aircraft began in the early 1900s. The first vertical takeoff and landing (VTOL) aircraft was built in 1924s, named Berliner No. 5. It reaches 15 ft height in 1 minute and 35 seconds. The evolution of flying cars is shown in the Figure 2.1. Then from the 1950s to the late 1980s, several companies provide early UAM services using helicopters in major US cities. However, safety and fuel costs create challenges for mainstreaming. Advancements in technology, sensors, and high-performance infomercials opened a new aura for unmanned aerial vehicles. Technological advancement, especially in electric propulsion and battery storage, led to various flying concepts.

Cloud computing is the provision and availability on demand of information technology services and resources such as computing power, storage, databases, analytics, intelligence, software, and networking via the Internet to achieve economies of scale as a result of pay-as-you-go pricing models. Cloud computing is also known as utility computing. You will be able to reduce the costs of your operations and manage the infrastructure of your company more easily and effectively as a result of this. Cloud service providers (CSPs) are the names given to the businesses that offer these types of services.

2.2 Requirement for Urban Aviation

2.2.1 UAM vehicle concepts and classification

The inability of the current transportation system, which is also called the ATM (air traffic management) system, which manages transportation, is the primary inhibitor of the development of urban air transportation [3]. ICAO classifies airspace using seven classes, namely alphabets A−G; the first one is controlled airspace that includes classes (A−D), [2] and the other one is uncontrolled airspace that covers the remaining classes (Figure 2.2) [4]. Every class of airspace is a set of protocols that describe how the airspace

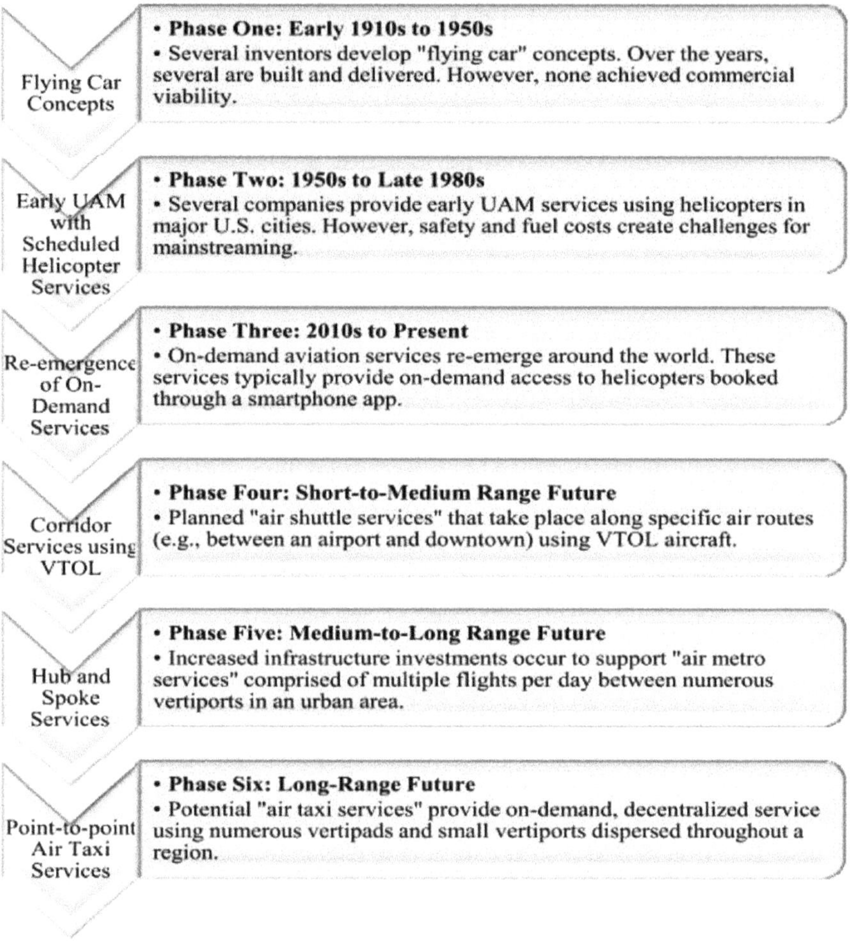

Flying Car Concepts	• **Phase One: Early 1910s to 1950s** • Several inventors develop "flying car" concepts. Over the years, several are built and delivered. However, none achieved commercial viability.
Early UAM with Scheduled Helicopter Services	• **Phase Two: 1950s to Late 1980s** • Several companies provide early UAM services using helicopters in major U.S. cities. However, safety and fuel costs create challenges for mainstreaming.
Re-emergence of On-Demand Services	• **Phase Three: 2010s to Present** • On-demand aviation services re-emerge around the world. These services typically provide on-demand access to helicopters booked through a smartphone app.
Corridor Services using VTOL	• **Phase Four: Short-to-Medium Range Future** • Planned "air shuttle services" that take place along specific air routes (e.g., between an airport and downtown) using VTOL aircraft.
Hub and Spoke Services	• **Phase Five: Medium-to-Long Range Future** • Increased infrastructure investments occur to support "air metro services" comprised of multiple flights per day between numerous vertiports in an urban area.
Point-to-point Air Taxi Services	• **Phase Six: Long-Range Future** • Potential "air taxi services" provide on-demand, decentralized service using numerous vertipads and small vertiports dispersed throughout a region.

Figure 2.1 Phases of flying cars.

will glide and how the in-flight traffic flow controller interacts with those aircraft. Overall, this allows a country to better control the flow of planes within its airspace for safety and security. Now that the seven classes have been created and defined, the ICAO stands for the international civil aviation organization. These were created to standardize the airspace of It's the Planet all countries adhere to these regulations. However, they do alter the airspace classes to suit their own needs. For example, a country does not have to use all seven classes to control their airspace; they may only use maybe four out of the seven classes, and a country can also add specific rules to their

Airspace Classes

	Flights	Separation	Traffic Advisory
Class A	IFR Only	All Flights	N/A
Class B	IFR + VFR	All Flights	N/A
Class C	IFR + VFR	IFR > IFR/VFR VFR > IFR	Provided to all VFR flights
Class D	IFR + VFR	IFR > IFR	IFR > VFR VFR > IFR/VFR
Class E	IFR + VFR (VFR flights do not require clearance)	IFR > IFR	All flights (when practical or possible)
Class F	IFR + VFR (No clearance needed)	IFR > IFR (when possible)	All flights, if requested (when practical or possible)
Class G	IFR + VFR (No clearance needed)	Not provided	All flights, if requested (when practical or possible)

Figure 2.2 Airspace classes of urban air mobility.

airspace regulations. Now let us explain the basic differences between the seven classes as defined by the ICAO.

The seven classes fall into two main categories of airspace: controlled and uncontrolled. So, in the first one, airspace is an area of the sky, where air traffic control (ATC) has authority; they give instructions like cleared-for-takeoff and cleared-to-land. The ATC are in charge, and they are controlling the planes. Uncontrolled airspace is just the opposite of controlled airspace. Air traffic control is present in uncontrolled airspace, but there is more of an advisory service; they can provide weather updates, for example, if a pilot requests them [5]. The ICAO dignified every airspace type of flight. The first one is IFR, which stands for instrumental flight rules and another is VFR, which stands for visual flight rules. The ATS service gives an insight into what is being done inside the loop; it takes care of communication between different flights and their clearances, VFR flights, their travel information, and their operation information as well as speed limitation and altitude [4].

In class A, the airspace that is allowed is for IFR flights, and they must be cleared by air traffic control. Air traffic control is responsible for the safety of different flights by safeguarding different planes with distance, so that they will be at a safer distance from each other. In class B airspace, both are allowed to provide the difference and to give the clearance. In class C airspace, things start to get a little complicated. First, IFR and VFR

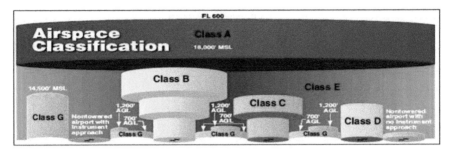

Figure 2.3 Airspace classification.

flights are allowed, but here is the difference. Air traffic control delivers parting and different protocols to keep IFR flights away from IFR and also VFR. Our objective is to distinguish VFR (visual flight rules) flights from IFR (instrumental flight rules) flights. Instead of providing traffic separation directly, VFR pilots will receive a traffic information service, where they will be informed about other VFR traffic by air traffic controllers. Nevertheless, it remains the responsibility of the pilots involved to maintain their separation.

IFR flights including VFR are allowed in class D. Here also, air traffic control provides intelligence for IFR flights and provides various in-traffic information about VFR flights. Figure 2.3 illustrates the classification of airspaces. However, VFR flights are only allowed to fly in the airspace of class E, even without getting information about the clearance from ATC IFR. The main thing about flights in this class is that IFR flights are separated from each other there's nothing with VFR, and all aircraft will be given traffic information, but only whether it is practical, or whether it is possible, it is not a service that will be given all the time.

The fact that a variety of different standards, certifications, and operational authorizations are required by the country and its laws is one of the many factors that contribute to the difficulties that the UAM transportation system experiences as a direct result of policies and regulations. These difficulties are caused by several other factors as well. One has a responsibility to make certain that the activity in which they are engaging does not provide any dangers from the perspective of anything and everything.

Infrastructure: For a transportation company seeking to modernize its outdated system or improve its services, a strong emphasis on infrastructure is vital. The project must gain acceptance and support from the local community. Furthermore, it should prioritize safety regulations and environmental mitigation, while also ensuring robust physical and cybersecurity measures are in place.

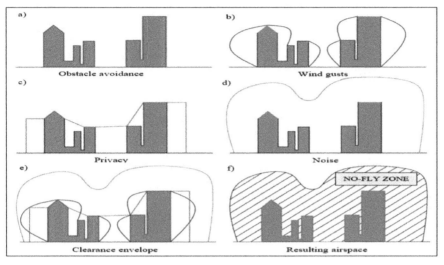

Figure 2.4   Space availability.

Let us apply reasoning to discover things that could impede mobility and alter the state of space available to flight. Figure 2.4 demonstrates the factors causes the availability of space.

The aspects are divided into four groups:

1. Physical and cyber security
2. Public aspects
3. Variables connected to the machine feature
4. Elements related to aircraft.

The advancement of distributed electric propulsion (DEP) and power-driven technology has sparked numerous developments in jet technology for urban air mobility (UAM) [1]. DEP offers innovative ideas and ample manufacturing possibilities, resulting in a wide array of vehicle forms due to varying primary functional requirements [2]. To classify and understand these vehicles better, a two-step categorization is employed, considering factors like revitalization during the journey and the machinery enabling vertical takeoff and landing (VTOL) capabilities [3].

Early VTOL ideas were influenced by rotorcraft, such as helicopters, which produced lifts with alternating horizontal rotors. Since the creation of VTOLs, several other designs have been created, including Cyclogyros [4], which produce lifts using cyclic rotors with a horizontal axis. These designs, sometimes known as wing rotors, feature characteristics of helicopters while simultaneously having the performance benefits of fixed-wing aircraft [5].

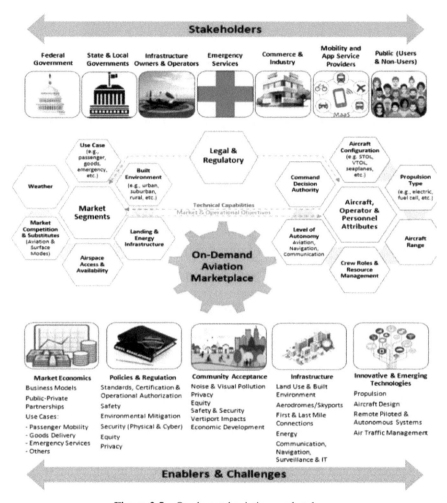

Figure 2.5 On-demand aviation marketplace.

Another VTOL design uses tilt rotors to create lift, while longitudinal propellers are mounted on axles at the extremities of fixed wings to propel and lift. Figure 2.6 clearly shows the UAM VTOL vehicles categories. The F-35 and Harrier are two of the most well-known military VTOL aircraft. Due to their ability to take off and land vertically and hover when there is not a good place to land, certain current drones can also be classified as VTOLs [6]. The power sources for VTOL vehicles include extra electricity, hybridization, and hydrogen fuel cells. Concepts of this group are much more efficient

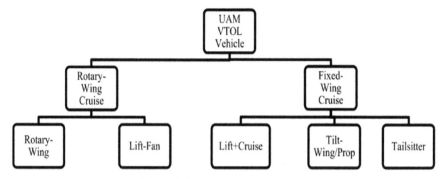

Figure 2.6   UAM VTOL vehicle.

and quicker during cruise flight than rotary-wing cruise configurations [7]. Integration aspects and the goal to avoid diminishing cruise efficiency limit the installation of large rotor areas required for effective hovering flight [8].

Most VTOLs can land on roofs, grassy areas, and meadows with little to no need for floor space, unlike conventional aircraft (CTOL). This has several benefits for the movement of both people and goods, and it works particularly well in crowded urban environments. The use of already-existing helipads, parking lots, retail malls, and other facilities might be temporary alternatives to the need for specialized substructures [9]. If VTOLs were solely dependent on inhabited areas for takeoff and landing, this would have the issue of preventing their operation. A full physical and digital structure will be needed to enable safe and effective operations because of the rising complexity brought on by widespread adoption

2.2.2 Operative perceptions, the structure of the market in addition to amalgamation toward the current transportation system

The UAM market may be segmented into the several sub-markets it contains as well as the various market participants. When various distinct ownership arrangements are being examined, the provision machinist, who is also the owner of the automobile, plays the most significant duty in integrating the UAM system. Figure 2.5 shows the On-demand aviation marketplace of UAM. This is because the provision machinist is also the owner of the car. Different market levels are being observed.

Platform provider: It is the means via which interaction with the end-user takes place, following which the UAM service is supplied in line with the outcomes of the engagement that took place through that channel. These vendors can be a company or a firm that offers a platform to organize and deliver services among various suppliers of such services and works as a mediator between them. In other words, they can operate as a vendor marketplace.

The second provider is the one who provides the transportation service for the overall business. This particular supplier is commonly referred to as the service provider. This is accomplished by scheduling flights, doing maintenance on the vehicle (including charging its fuel tank), keeping track of its schedule, as well as completing repair and overhaul (MRO) and cleaning. The company that provides transportation services for the entire enterprise can establish a connection with the close client, and it is also possible for it to provide the close customer's transportation facility through businesses that are peripheral to the company. Either option is open to the company that provides services on a global scale (platform).

Vehicle owner: The firm that owns the UAM cars may be the same company that provides the service. It may be a different company whose primary work is that he/she provides UAM vehicles. It might happen through a long-standing leasing arrangement or a temporary and lithe agreement.

Vehicle constructor: Presently, numerous firms are evolving UAM automobiles, which is relatively high when equated to, say, automobile, train, or airplane manufacturers. Other layers include ground infrastructure providers, and also those providers that give insurance and follow UTM providers, and insurance companies [10].

Still, we are not sure that UAM will provide public transit. Therefore, effective development with already-existing modes of transportation, particularly PT, which is public transportation, is crucial. Policymakers should review the financing model used by most transport offerings to allow UAM systems that complement PT rather than compete with it. Similar debates are presently going on in autonomous ground vehicles, which might have detrimental impacts on PT and other means of transportation. Researchers have highlighted physical, fare, service, platform integration, and data interchange as essential preconditions for successfully integrating autonomous ground.

When it comes to selecting a mode of transportation for UAM, other crucial factors to take into account are travel expenditures and journey time. As a result, it is essential to have a system that is both well-designed and capable of cutting down on transfer time. If the operator does not address the proper target segments, the project may come to an end as a result of the considerable travel expenditures that would be incurred during the first deployment of the system (Holden and Goel, 2016). Transparent service and technology are also essential for the user and public adoption, as is educating people about the steps taken to protect (data) privacy as well as the safety and security of both system users and non-users. Both of these factors are crucial for the user and public adoption of the system (Al Haddad et al., 2020). It is possible that one tactic would be to get the ball rolling on early initiatives to raise public awareness regarding automation. The initial assessments on UAM market shares suggest large variations in prospective adoption rates since they utilize extremely diverse price assumptions. It is estimated that the modal share is somewhere about 4%, and UAM costs at taxi pricing levels converge around that number.

2.3 Air Taxi Services (ATS)

As a result of a variety of factors, including high anthropology and an increase in privately owned automobiles, as well as an increase in commuters coming from less populated regions, who are mostly contributing to transportation as passengers, traffic becomes a significant issue in metropolitan areas. In major cities where the typical traveler faces at most 90 minutes in traffic, tension and anxiety rise [11]. Three hundred and thirty grams of CO_2 are released into the environment each mile due to these congestions. With the increase in pollution, it will degrade the environment and the health of passengers too, which will also increase the expenditure of ones over his/her medication.

Let us take an example. Manhattan loses $20 billion annually because of road traffic, with additional petroleum and vehicle running costs making up approximately 13% of the whole loss [15]. Therefore, it is crucial to look beyond ground transportation options to speed up daily commutes for those living in cities and reduce traffic. There were innovations like flying taxi services being developed by several logistics businesses and agencies to capitalize on urban air mobility (UAM), and there was also one more innovation: the aviation ride-hailing concept that works on the same principle. Aircraft companies have been actively engaged in the production of air taxis over the last several years. In addition to Uber, which anticipates launching its flying taxis (known as Uber Elevate) in 2023, ATS, which stands for air taxi service, might provide a quicker and more dependable means of transportation by using the VTOL vehicles suggested [15]. In addition, electric air taxis are projected with much more environmental friendliness, security, silence, and futuristic features than any current means of transportation [15].

2.4 Urban Air Taxi Services (UATS)

ATS is parallel to typical means of transportation. Requesting an ATS would be analogous to making a reservation for a typical ODM service. ODM service means on-demand mobility. A registered consumer will use a ridesharing application that enables the user for the pickup and drop-off locations. The platform will calculate the rate and journey time for all eligible means of transportation, including ordinary taxis and flights, based on the trip details [16]. If an ATS is practical, a client may use the service based on several factors, including their desire to travel, the cost of their journey, and how sensitive they are to transit times.

An ATS (air taxi service) is projected to have several innovative ideas and different segments. It works the same as the day-to-day taxi service where a taxi comes to you and you tell them the drop-off location, and then you travel to that place, but here the difference is that it is autonomous and has innovative features. The workflow of the air taxi service compared with on-road is demonstrated in the Figure 2.7. The initial journey travels from the pickup site to a vertiport via an on-road vehicle [4]. And then the whole process is completed by a taxi that flies the commuter from the starting sky-port to the

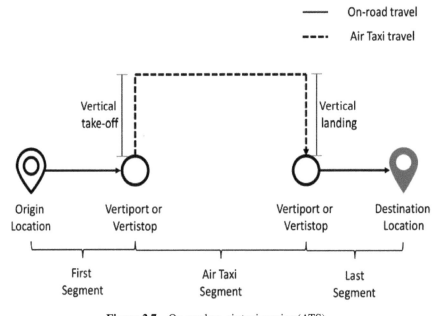

Figure 2.7 On-road vs. air taxi service (ATS).

Figure 2.8 Air taxi image.

destination. However, the client could be instructed to walk if the journey's initial or end leg is near the specified sky-port [15]. Air taxi services as shown in Figure 2.8 are now widely accommodated by many cities.

Air taxi services are now widely accommodated by many cities. An example is Boeing and Wisk.

Recently, Boeing and Wisk's plan for air transportation of passengers and freight has been accessible. According to this concept, autonomous and unmanned aircraft will soon be able to transport people and freight around cities and suburbs. Businesses informed the broader public of their plans.

The road map illustrates the most effective way to implement urban air mobility (UAM) in the United States and provides ideas on how to do it at the technical, governmental, and social levels. In the United States, urban air travel must become routine. In addition, the optimum technique for incorporating UAM into the national airspace is discussed. It will be possible to automate flights to relieve controllers and pilots of some of their responsibilities. To accomplish this objective, we shall accept unmanned flights at any time of day or night as long as they conform to visual or instrument flying requirements. To do this, flying robots will be necessary. With an automated infrastructure, these technological advances may be accomplished on the ground and in the air.

According to Gary Gysin, CEO of Wisk, the considerable effort we are doing at the moment is a stepping stone for the growth of UAM in the United States and internationally. We are sharing it immediately since we consider its importance. We are happy to provide you with access to this information at this time. At the beginning of this year, Wisk revealed plans to build an electric vertical takeoff and landing (eVTOL) air taxi for four passengers. Without a pilot, people will be able to fly to large cities on the aircraft of the sixth generation, which is now undergoing testing. In this circumstance, the aircraft will be able to continue flying without aid from the outside world.

In addition to offering more room for people and their bags, the reconfigured seating makes it easier to help individuals who may have difficulty traversing the airport. Due to the aircraft's increased space, it may be used for numerous purposes. This increased capacity gives the airplane greater space for passengers and the ability to transport more cargo, allowing it to better serve new applications and use cases. This may be accomplished by increasing the size of the grip.

The vision we have built, according to Gysin, is the result of years of cooperation with significant organizations like NASA, the FAA, and Boeing. Several market players further affected this concept. The authors of the study assert that as a direct result of its findings, the research provides the most complete framework and a picture of how UAM may be integrated into the nation's airspace in the future. They have studied the results in a way that bolsters this claim. Boeing and Wisk will build new infrastructure, such as vertiports, as part of the architecture that is being constructed. This is what "the aim we have set" alludes to UAM aircraft.

Electric vertical takeoff and landing (eVTOL) aircraft were able to land on vertiports. Numerous multinational aircraft manufacturers adopt this innovative method. This proposal was made as a possible solution for any problems that cargo electric vertical takeoff and landing (eVTOL) aircraft and electric vertical takeoff and landing (eVTOL) air taxis may encounter in the near or far future. Supernal and urban-airport, two subsidiaries of the air taxi company Hyundai, have developed an infrastructure center with a range of transportation alternatives this year. This endeavor will take place at the Urban Airport. During a natural disaster, these portable vertiports may be moved to support emergency services or offer shelter for commuters, depending on what is necessary at the moment.

Moreover, both Wisk and Boeing agree that constructing a fleet operations center in the future would be a wise decision. In this facility, airplane surveillance would be the responsibility of supervisors of many vehicles.

They would also be responsible for following air traffic control directions to maintain separation between planes and ensure the safety of the flight.

2.5 UAM Modeling Approaches

Rothfeld et al. (2019) highlight the foundations for the introduction of UAM into transport simulation frameworks in the present setting. There have been some studies done on the requirements for the infrastructure or the operating constraints. An inquiry brought up the possibility of developing and utilizing a UAM extension for MATSim; as a result, this extension has been outlined in the initial set of tests. MATSim is easily expandable and is agent-based thanks to these features, which are essential for individualistic transportation modes because they make it simpler to integrate UAM in transport models and permit user-centric outcome evaluations. Individualistic transportation modes are characterized by a high degree of autonomy and personalization.

When paired with traditional modeling frameworks for UAM, the results of several studies that were carried out in the past have proved the promise that is presented by on-demand UAM as a new development in the transportation industry. In addition to this, as can be seen in the figures that are presented below, they demonstrated the impact that the primary operating

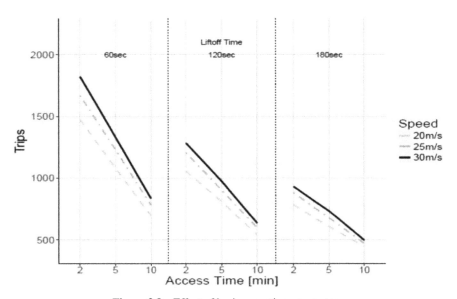

Figure 2.9 Effect of basic operating parameters.

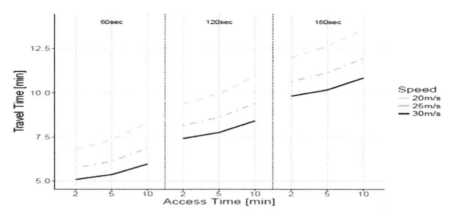

Figure 2.10　Effect of basic operating parameters.

parameters had. The various UAMs all have different access times, lift-off times, and cruising speeds, among other differences.

The effect of the operational UAM attributes stated earlier on the total journeys is shown in Figures 2.9 and 2.10, which displays the time taken by various customers who use the same UAM characteristics for variants. Combined, both charts demonstrate the negative correlation between falling demand and a means of transportation where the trip durations are growing, provided those other elements, like price, stay the same.

It is something that commonly takes place as a result of the ground-based transportation tactics that are utilized in this day and age. Customers were able to directly use UAM from both the point of origin and the point of destination of that agent thanks to the various UAM timings, but the length of any potential UAM customer trip that took place over a period that was measured in increments of time was lengthened as a result. The ability to specify UAM was made feasible thanks to the work of Rothfeld et al. (2018b, 2018a), which contributed to these simulation capabilities. These capacities were improved in several ways, and this was one of them. For VTOL operations that need prospective UAM passengers for entrance and egress legs, separate aircraft flights, their respective networks, and diversified stations located at a variety of sites or VTOL architecture are required. Figure 2.11 illustrates the structural components of a MATSim UAM station (Rothfeld et al. 2018b).

It links the ground transportation network (vehicle) to a clumsy representation of a flying network (UAM), which acts as a suggested UAM

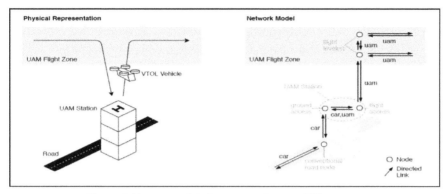

Figure 2.11 Network infrastructure for a MATSim UAM.

station in person. On the other side, Vascik and Hansman (2017) focus on the operational restrictions associated with UAM missions, more especially the problem of expanding the number of UAM flight phases. In the process of station placement, some of the elements that were taken into consideration were the job density, the cost of renting out office space, and the distribution of the area's median family income. Despite the use of coupling factor compeering in research, the use of a Delphi workshop, and the aid of an expert, Fadhil (2018) was unable to achieve a standardized method for a station that was acceptable for UAM. This was the case even though they had the support of an expert.

Successful UAM depends on a wide range of criteria that must be considered, yet many issues remain unanswered in the body of extant research. There is a dearth of literature on UAM's potential effects. Who benefits from the service? Evaluating how UAM improves rural access and its impact on social equality, including who benefits and who might be disadvantaged. Additionally, there are no talks on the system's inclusivity, particularly concerning those with special needs. First, discussions on modal changes and complementing public transportation rather than competing with it have just begun.

2.6 Modeling of Urban Transportation Systems

2.6.1 System dynamics model of urban transportation system

An essential component of creating a resource-efficient, environmentally conscious, and people-centered society is the sustainable improvement of transportation systems in urban areas. The definition of sustainable trans-portation includes four components: transportation sustainability, social sustainability, environmental sustainability, and economic sustainability.

The urban transportation system is an intermingled system with many variables, loops (feedback) between subsystems, and factors affecting the topology of transportation. The conventional linear quantitative technique should not be used to characterize the properties of this complicated system. As a result, this work system dynamics (SD) technique is employed to mimic the development of the urban transportation system.

Forrester from MIT created the SD in the 1950s. The method is employed in quantitative studies of the complex socioeconomics sector and is defined by the feedback control theory, coupled with computer simulation technology. Equations, variables, and feedback loops are used to implement SD techniques. The variables are as follows:

- level variable (defines a flow over continuous periods);
- rate variable (defines a flow over a specific period);
- auxiliary variable (it identifies the rate variables).

Equations in integral, differential, or other forms connect the three main types of variables. The complexity and breadth of the urban transportation system make it unsuitable for simulating and studying it using conventional methods. Therefore, complex system analysis has used the SD technique. The first SD model was put out by Forrester and is used to model urban systems. The SD technique is often used to assess the effectiveness of local, sustainable development, examine the connection between land use and transportation, and calculate the environmental effects of industrial gardens.

Three steps may be identified when adopting an SD approach: prelimi-nary, defined, and thorough analysis. Figure 2.12 shows the phases of system

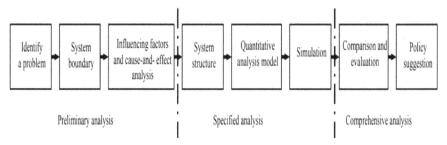

Figure 2.12   Three phases of the system dynamics (SD) technique.

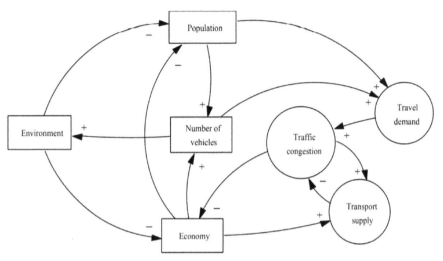

Figure 2.13 Barriers to urban transportation.

dynamics. In the early analysis, it is vital to specify the external and internal variables, particularly the casual feedback loops of the variables, as the knowledge of the system features deepens. Based on the findings of the preliminary analysis, the system structure is built in the defined analysis, and coefficients and equations are provided to carry out a quantitative simulation process.

The population sub-model, economics sub-model, and vehicle number sub-model are crucial to quantitative analysis and significantly impact the environment and transportation systems. The interplay between the transportation demand and supply sub-models and feedback to the economic development sub-model leads to the congestion sub-model. As shown in the Figure 2.13. Both the population sub-model and the economics sub-model are impacted by the environment sub-model, representing the main barrier to the growth of urban transportation.

2.7 Intelligent Transportation System Modeling

Due to the heavy traffic and congestion on roads, the ITS modeling helps a lot in each aspect. The ITS modeling deals with two principle: the management of travel and transportation demand and the supply. This modeling refers to the traffic smoothness in the transportation system. These two topics are interconnected.

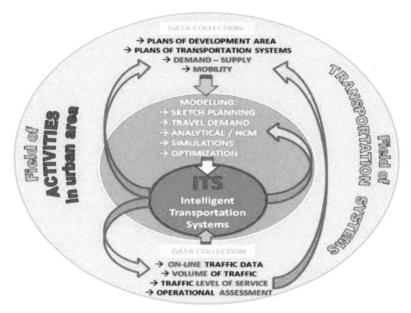

Figure 2.14 Four-stage transportation models.

The stages of transportation model is shown in the Figure 2.14. The ITS services are helpful tools for managing transportation. The efficient system includes modifications to the city's transportation infrastructure and its environs and an appropriate impact on choices made by its users about the route and timing of planned trips. The usage of several ITS service packages is necessary due to the vast variance in system users' travel patterns and destinations.

The ITS services have a significant role in influencing the intended travel behavior. Within the framework of ITS, traffic management and control may be effectively managed and controlled using transportation system modeling. Practical applications of transportation models include: evaluating, simulating, or optimizing the working of transportation; modeling current working and predicting the various outcomes that we get and using those outcomes for the betterment of the structure of innovation; assessing different contexts, such as designing, working on infrastructure, and enhancement in technology.

Four-stage transportation models, the most often used travel demand models, enable projecting the future found on present circumstances and futuristic estimates of family and work characteristics. These models account for the choice of destination, mode of transportation, time of day, route,

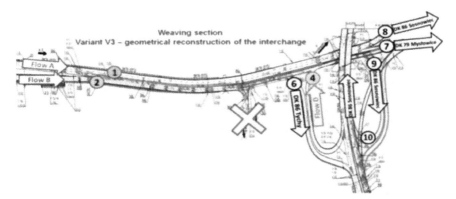

Figure 2.15 Traffic movement in the ITS systems.

and depiction of rush-hour traffic [17]. Tactlessly, due to their meager representation of the dynamic nature of traffic and their lack of ability to sense various things like speed, flows, and density, these models are inadequate for evaluating the effects of ITS systems on transportation systems.

The three models namely meso, macro, and microscopic simulation models provide a better option for assessing the effect of ITS. ITS systems are data sources (with issues related to real-time big data processing and exploration).

The ITS services provide technology options for collecting the data, such as sensors and recorders to monitor traffic volume and parking space availability. Internet, LAN, WAN, and various other technologies such as AI and IoT guarantee to deliver data to users. Additionally, transportation models are required for ITS systems to operate to their full.

The delivery of ITS services for transportation is accomplished via interconnected systems known as intelligent transportation systems (ITS). The functions necessary for ITS are the physical entities such as the field, data, and information drifts together, which operate and result in a better system. The information exchanged between ITS system elements is all defined by an ITS architecture [12].

The ITS systems mentioned above can enhance how smoothly traffic moves through the user services bundled into service packages. Figure 2.15 illustrates the traffic movements in the ITS systems. As a result, the subsequent picture of ITS protocol and its working is necessary for modeling projected traffic smoothness that considers ITS user services and service packages. Figure 2.16 represents the model for smooth working in urban transportation.

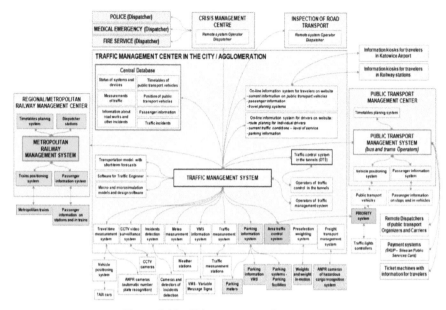

Figure 2.16 Urban transportation systems.

2.8 Multiplex Networks

Multiple transportation modalities that are coupled together make up public urban mobility networks. The multi-layer character of transportation networks is often ignored in most urban mobility and planning studies, which only consider aggregated representations of this complicated situation. So, we will discuss a model showing a city's whole transportation system as a multiplex network. Service schedules and waiting times can be viewed in two ways within the multiplex network. One perspective treats each transit line as a distinct layer, representing individual models. The other perspective groups these lines to form a unified transportation model, enabling the replication of real-world data [18].

Although contemporary research on multiplex networks has mostly disregarded the sociotechnical side of urban transportation systems, the multimodal component of these systems has attracted a lot of attention. However, this is the situation. However, there has been a lot of interest recently in the multimodal aspect of urban transportation networks. Urban transportation networks have emphasized multimodality and sociotechnical integration in recent years. The creation of tools for the analysis of the impact of newly added transportation layers on the current dynamics, as well as the proof that demand structure, plays a significant quantitative and qualitative role in the equation, which take up a significant section of this work [13].

Multiplex networks are helpful metaphors for systems where many interactions may link the same group of nodes. Multiplex systems can be described with a few example social networks, the system on which transportation is carried on with many modes of transit, and biological systems that consider various sorts of interactions [18]. Nodes and connections in a multiplex network are organized into layers based on their use. Layers include information stolen if we would consider equivalent conglomerate networks since they might be interconnected. Additionally, it has recently been shown that various dynamics applied through the utmost of multi-layer connection deliver fresh perspectives on the studied issues. Few studies have been conducted on multiplex networks on the working of transportation in urban areas to examine their multimodal behavior and coupling or resilience. In a sense, a super layer is created by grouping all the lines that operate in the same mode: buses, metro, and trams. This depiction is relatively thin − only a few layers − but it completely ignores waiting and transfer periods across lines of the same model, which might potentially result in erroneous estimates of journey durations or shortest pathways. An alternative is to treat a particular line in each mode as a separate layer. Even if, in the instance, we can manage the time between the transportation and the synchronization among stations on various layers, it is impossible to gauge the significance of a particular form of transportation for system mobility [13].

Multiplex network analysis is a common practice that serves as a core tool for gaining a system-level understanding of multimodal transportation networks. The topology and geometry of the transportation infrastructure have always been the main research topics in any study that has ever been done with a focus on transportation systems. For instance, they have looked at the connections between various transportation networks, some of which are much larger than others. These networks include commuter trains, bus routes, subway systems, and air traffic. Due to the sociotechnical character of transportation systems, the interplay between infrastructure and user behavior − and, more particularly, the organized pattern of user demand − determines how these systems behave. This shows that the interaction between infrastructure and user behavior may alter how transportation networks operate. The behavior of transportation systems is influenced by how users' behavior interacts with their infrastructure. Transportation systems' behavior is governed by this relationship.

The number of persons who travel between each potential origin and destination in the city is included in each of the cells of the origin−destination (OD) matrix, which is a popular method for characterizing this demand. Each item in the list gives the total number of passengers for a specific city's starting point and ending point combination. The number of people passing through a city between any two places that are designated as their starting and ending points may be counted. Nobody realized it was possible to make OD matrices until very recently. The great majority of urban transportation system models in use today use *a priori* assumptions about the OD matrix's composition. This is due to the lack of information upon which to build these models. When there is a dearth of understanding, these assumptions are made. We now have access to a wealth of data, making data-driven, scalable sociotechnical analyses of urban transportation systems possible. This is because we have access to a vast amount of data [14].

Comparing the results of this research to those of the uniform and stochastic OD models shows two more levels of realism. One may be able to arrive at more accurate conclusions regarding system-level indicators like distributions of commute times by starting with the most basic urban geometry concepts. One example of this idea is the length of time needed to go from one place to another. People who travel to work, for instance, often live far closer to their places of employment than one could expect based only on the principles of random chance. Second, it is easy to foresee that the network would become crowded since exact data on the traffic demand is readily available [15]. If one wants to have precise trip-time estimates for routes that use crowded or underutilized network connections, it is essential to gather the relevant data and keep track of it. This is because it has a considerable effect on the accuracy of journey-time estimations. However, multimodal research has merely speculated about congestion, even though it is a crucial part of the contemporary transportation system.

2.8.1 Techniques and resources

The researchers used call detail record (CDR) data to create origin − destination (OD) matrices for the network. They used Riyadh's proposed metro as a case study. In the neighborhood, 5.8 million people are living. The peak morning commute flows, which occur between 7:30 and 8:30 in the morning, are accurately depicted by the OD matrix that was employed. As a result, exact network simulations could be performed at the required time intervals. They employed an iterative traffic assignment approach to distribute traffic in a way that was compliant with the OD matrix. Twenty percent of the flows were caused by the top four demographic groups that make up the commuter population, ten percent by the next two groups, and so on. We chose to adopt this grouping instead of the more popular (40%, 30%, 20%, and 10%) since doing so would have prevented us from achieving high metro speeds due to an excessively high level of congestion at metro stations [16]. We would not have been able to go at high metro speeds because of the congestion. According to their particular preferences, the top 20% of users were given the option to choose the network routes that offered the fastest free-flow journey times possible. The travel times for each sector were then adjusted using the regular BPR technique to account for the effects that congestion would have on them.

$$t_e^* = t_e \left(1 + A \left(\frac{j_e}{c_e} \right)^B \right).$$

The symbol j_e shows how it flows during this iteration, while the symbol t_e shows how much time it has to flow freely. The letter c_e shows how much water can flow through segment e, while the letter j_e shows how much water is moving through it right now. For this chapter, we will use the default values, which are 0.15 and 4, respectively [13]. A and B are both thought of as variables because their values can change. When moving through certain parts of the metro system, the amount of time spent in transit is always the opposite of how fast free flow can happen. This is because people often think that not enough people use the metro. Then, based on these newly calculated edge costs, we let the next group choose its way through the network. After that, the next group was free to choose its path through the network. It took a lot of time in this process to make sure that every group was formed.

We were able to figure out how the dynamics of the multiplex depend on the metro layer by doing the above task at different metro speeds and comparing the results. For the sake of argument, let us say that the metro runs as a network at a mean speed of vc, which is a free-floating parameter, and $vc = 38$ km/hr is the average speed of traffic on the roads during rush hour when there is no metro. This will give us the chance to move on to the next step. There is a small chance that metros with lower travel values move faster than those with higher travel values. When the value is set to 0.5, for example, the metro goes at a speed of 76 km/h [13]. When there is no metro, this is about the same as moving twice as fast as on the street level. We were able to figure out that the average effective metro speed in Riyadh is 47 km/h by combining data from metro systems that had already been measured with information about how the city's technology works. Using this information, we were able to figure out that 38 km/h divided by 47 km/h is about 0.8.

2.8.2 Flows that have been assigned are validated

The case study evaluated the precision of our traffic modeling by comparing our most precise travel time predictions with the data provided by Google Maps. We used a web crawler to request Google Maps to get journey times for the top 679,085 OD pairs in terms of flow volume. This request included both periods when there was a smooth flow of traffic and times when there was a bottleneck. The chart in Figure 2.17 shows a comparison of the travel times predicted by Google and ITA for the same journey. After that, we adjusted the speeds of the free-flow streets so that they were more comparable to the estimations provided by Google Maps for the amount of time it would take to travel without encountering any traffic, and we incorporated these updated speeds in our calculations moving forward. After that, we proceeded to the subsequent series of computations using these revised speeds. Although the match was better for estimating open flow than for projecting choked flow, we discovered that both were adequate for modeling purposes. However, the match was better for estimating open flow [17]. Even though estimations of free flow now have a higher degree of precision, this occurred.

The map in Figure 2.18 depicts the flows that occur throughout a large spectrum of diverse values of. After the ITA had finished designating routes,

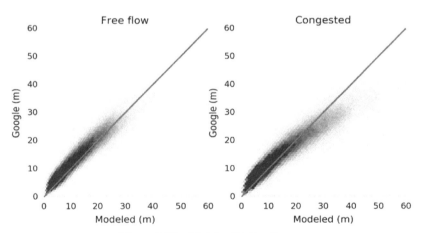

Figure 2.17   Model validation flows.

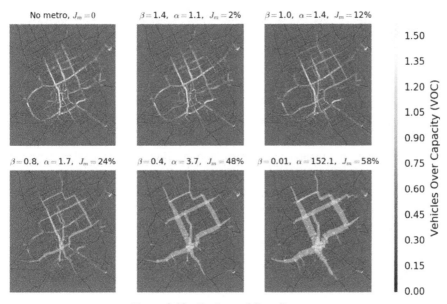

No metro, $J_m = 0$ $\beta = 1.4$, $\alpha = 1.1$, $J_m = 2\%$ $\beta = 1.0$, $\alpha = 1.4$, $J_m = 12\%$

$\beta = 0.8$, $\alpha = 1.7$, $J_m = 24\%$ $\beta = 0.4$, $\alpha = 3.7$, $J_m = 48\%$ $\beta = 0.01$, $\alpha = 152.1$, $J_m = 58\%$

Vehicles Over Capacity (VOC)

1.50
1.35
1.20
1.05
0.90
0.75
0.60
0.45
0.30
0.15
0.00

Figure 2.18 Designated flow diagram.

the next step was to determine the empirical speed ratio of the subway to the streets at each level. Although it is a system statistic that indicates real congestion speeds, the (mean) ratio of "posted speed limits" is used as a component in modeling. Even though it is used, this is carried out. Also necessary is an analysis of the projected subway system's feasibility for passengers under actual traffic circumstances [18]. Since the metro is slightly quicker than the crowded street layer, high population density areas do not need inhabitants to relocate. The calculated practical speed ratio of 0.8 is similar to the observed speed ratio of 1.7. This is because the metro travels faster than the clogged streets, which has resulted in a large rise in the number of people using its service. The subway becomes less crowded as more people use it. It has been observed that numerous important thoroughfares are now more crowded than they were before to the building of the metro, even though there is less traffic overall [17].

Even when there is just a little quantity of traffic, the metro system can effectively handle 58% of the total flow. When compared to the highway network, the metro's coverage area is far smaller, and this distribution restriction remains even when metro speeds are raised. Forty-two percent of all traffic consists of individuals traveling to or from metro stations or taking diversions that need more time than driving straight to the destination. These travels constitute the majority of flow-related journeys.

The movement is shown in the following graphic (Figure 2.19), along with what occurs when metro speeds are increased. Although estimated to be $J_m = 58\%$, the asymptotic limit of this expansion is still getting closer. The number of people who often use the subway and the speed difference between it and the street layer are significantly correlated. It is quite difficult to estimate how this would affect travel times. The figures show that when the metro's speed rises, the usual travel times decrease linearly and reach at least 43% of their initial values. Compared to its original value, the standard deviation of travel durations has decreased by at least 43%. In contrast, driving time quickly reverts to its pre-reduced level: at = 0.6, 33% sooner than our logically projected = 0.8, more than 85% of the whole driving time decrease is realized. The pace at which the total amount of time spent in congestion will decrease is also represented by 0.6. The moment at which something happened has passed. These two factors are essential for reducing the detrimental effects of urban mobility on the neighborhood and city residents [17].

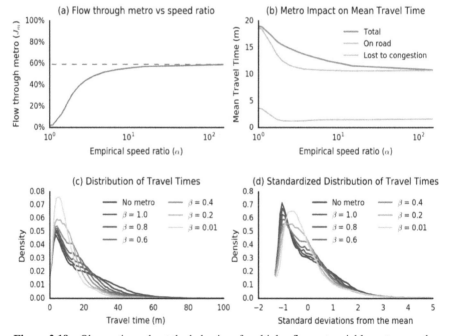

Figure 2.19   Observations about the behavior of multiplex flows at variable metro speeds.

According to the analysis, a little increase in metro speed beyond what is currently planned may benefit the environment. The current strategy calls for maintaining the current pace, which supports this conclusion. These results also imply that further speed increases may result in a decrease in the environmental advantages. A transportation model that is both more precise and extensive than those that came before it is necessary to properly analyze the pace at which consumer returns start to fall. The speed of the metro affects both the range of likely completion durations as well as the median amount of time needed to complete a journey. The quick metro's introduction, which dramatically changes the spatial arrangement of the urban transportation network, is to blame for this shift in quality. Formerly physically distinct areas have been reunited to achieve this goal.

2.9 Summary

With such a rate of urbanization, mobility is large. Metropolitan regions face important challenges. Traffic capacity is among the most recognized problem in Metropolis. The need for mobility exceeds the road and underground capacity daily, and, as a result, transport saturates during peak hours with a drop in mobility. Maintaining a good traveling speed for cars makes the urban environment particularly dangerous, especially when road traffic interacts with cyclists and pedestrians. Safety remains a challenge in the urban center movie, surface mobility, and internal combustion. Engines are responsible for aggravating air pollution in large cities, and the limits suggested by the Public Health Organization are frequently exceeded. This is a clear signal that sustainable development of urbanization cannot be based on the massive use of cars; certainly not with internal combustion. Finally, the large amount of people in need of transport also involves labor. Autonomous vehicles are now becoming more common, using intelligent systems to navigate the wide range of situations encountered in surface transportation.

Significant challenges exist in ensuring the safe deployment of autopilot systems. The use of airspace above densely populated areas is already routinely taken up by helicopters, emergency services, like ambulances, and police intervention. Make use of helicopters in a big Metropolis. New York and São Paulo, in addition to the deployment of vehicles that fly electrically yet unmanned, has begun drone services, for reconnaissance packages, and delivery, and recently the Delft University of Technology has demonstrated the use of a drone ambulance to transport.

So finally in this chapter, we discuss the concept of airspace, its strengths and weaknesses, and a well-defined approach to designing urban airspace. What is the requirement for urban aviation and how the UAM market is divided into sub-market and market actors? Finally, this chapter discusses a few UAM modeling approaches. ITS modeling deals with two principle: the management of travel and transportation demand and the supply and a multiplex network. As a future perspective and sustainable environment, one must need a new perspective towards the analysis of urban air mobility and make enhancements including the perspective of urban planning.

References

[1] "Positioning Helicopters in the Urban Air Mobility Ecosystem," and-europe.org, Accessed: Aug. 01, 2022. [Online]. Available: https://asd-eu rope.org/positioning-helicopters-in-the-urban-air-mobility-ecosystem

[2] "The long-run effects of urban air mobility Straubinger, Anna".

[3] P. D. Vascik and R. John Hansman, "Constraint identification in on-demand mobility for aviation through an exploratory case study of Los Angeles," 17th AIAA Aviat. Technol. Integr. Oper. Conf. 2017, 2017, doi:10.2514/6.2017-3083.

[4] "Seitewurdenichtgefunden. - CycloTech." https://www.cyclotech.at/app lications/(accessedOct.02,2022).

[5] "VTOL: how does vertical take-off and landing technology work | WIRED UK." https://www.wired.co.uk/article/vtol-vertical-take-o ff-landing-explained(accessedOct.02,2022).

[6] "A Brief History of Vertical Take-Off and Landing." https://www.popu larmechanics.com/flight/a15072284/video-history-of-vtol/(accessedOc t.02,2022).

[7] R. Rothfeld, M. Balac, and C. Antoniou, "Modelling and evaluating urban air mobility - An early research approach," Transp. Res. Procedia, vol. 41, pp. 41–44, 2019, doi: 10.1016/j.trpro.2019.09.007.

[8] A. Straubinger, R. Rothfeld, M. Shamiyeh, K. D. Büchter, J. Kaiser, and K. O. Plötner, "An overview of current research and develop-ments in urban air mobility – Setting the scene for UAM introduc-tion," J. Air Transp. Manag., vol. 87, p. 101852, Aug. 2020, doi: 10.1016/J.JAIRTRAMAN.2020.101852.

[9] "Vertical Take Off and Landing (VTOL) Aircraft with Vectored Thrust for Control and Continuously Variable Pitch Attitude in Hover | T2 Portal." https://technology.nasa.gov/patent/LAR-TOPS-283(access edOct.02,2022).

[10] "Strategic Decision Factors for Key Innovation Leaders in Undefined High-Tech Market Environments Evidence from the Urban Air Mobility Industry," 2021.

[11] S. Rajendran and S. Srinivas, "Air taxi service for urban mobility: A critical review of recent developments, future challenges, and opportunities," Transp. Res. Part E Logist. Transp. Rev., vol. 143, Feb. 2021, doi: 10.1016/j.tre.2020.102090.

[12] G. Karoń and R. Zochowska, "MODELLING OF EXPECTED TRAFFIC SMOOTHNESS IN URBAN TRANSPORTATION SYSTEMS FOR ITS SOLUTIONS," Arch. Transp., vol. 33, no. 1, pp. 33–45, 2015, doi: 10.5604/08669546.1160925.

[13] G. Sun, J. Zhao, C. Webster, and H. Lin, "New metro system and active travel: A natural experiment," Environ. Int., vol. 138, May 2020, doi: 10.1016/j.envint.2020.105605.

[14] P. S. Chodrow, Z. Al-Awwad, S. Jiang, and M. C. González, "Demand and Congestion in Multiplex Transportation Networks," PLoS One, vol. 11, no. 9, Sep. 2016, doi: 10.1371/JOURNAL.PONE.0161738.

[15] S. V. Sharif, P. H. Moshfegh, M. A. Morshedi, and H. Kashani, "Modeling the impact of mitigation policies in a pandemic: A system dynamics approach," Int. J. Disaster Risk Reduct., vol. 82, p. 103327, Nov. 2022, doi: 10.1016/J.IJDRR.2022.103327.

[16] K. Nellore and G. P. Hancke, "A Survey on Urban Traffic Management System Using Wireless Sensor Networks.," Sensors (Basel)., vol. 16, no. 2, p. 157, Jan. 2016, doi: 10.3390/s16020157.

[17] A. Solé-Ribalta, S. Gómez, and A. Arenas, "Congestion Induced by the Structure of Multiplex Networks.," Phys. Rev. Lett., vol. 116, no. 10, p. 108701, Mar. 2016, doi: 10.1103/PhysRevLett.116.108701.

[18] W.-B. Du, X.-L. Zhou, M. Jusup, and Z. Wang, "Physics of transportation: Towards optimal capacity using the multilayer network framework.," Sci. Rep., vol. 6, p. 19059, Jan. 2016, doi: 10.1038/srep19059.

3

System Dynamics Model of Urban Transportation System

V. Sakthivel[1], Sourav Patel Kurmi[1], P. Prakash[1], Sam Goundar[2], and Jueying Li[3]

[1]School of Computer Science and Engineering, Vellore Institute of Technology, Chennai, India
[2]Department of Information Technology, RMIT University, Vietnam
[3]Department of Computer Science and Engineering, Konkuk University, South Korea
E-mail: mvsakthi@gmail.com; souravkurmi5683@gmail.com; prakash.p@vit.ac.in; sam.goundar@rmit.edu; lijueying1108@gmail.com

Abstract

The complex urban transportation system includes commuters, transportation service providers, government agencies, and urban planners. System dynamics modeling is a powerful tool for understanding and analyzing the behavior of complex systems over time. Multiple feedback loops in the model capture the dynamic interactions between various components of the urban transportation system, such as traffic flow, public transportation, and private vehicle usage. To simulate the long-term behavior of the urban transportation system, the model takes into account factors such as population growth, urban development, changes in transportation infrastructure, and technological advancements. This chapter presents an urban transportation system dynamics model that captures the interactions of various stakeholders as well as their impact on the system's performance.

Keywords: Urban Transportation System, Dynamics model, Feedback Loop, Casual Loop, Public Transportation

3.1 System Dynamics Modeling

Cities are the hubs of complicated and potentially disruptive economic activity. Transport management and citywide accessibility are perennial issues [1]: issues include (a) parking and traffic management, (b) inadequate public transportation, (c) challenges for non-motorized transport, (d) the transport industry's effect on surrounding, the environment, and the fuel consumption, and (e) traffic congestion their safety. The need for transportation facilities will rise in tandem with the population. Rapid Transportation of people and products is crucial. Problems in the transportation system must be recognized and fixed to reach this objective [2].

As we already discussed, an essential component of creating a resource-efficient, environmentally conscious, and people-centered society is the sustainable improvement of transportation systems in urban areas. The definition of sustainable transportation includes four components: transportation sustainability, social sustainability, environmental sustainability, and economic sustainability [3].

The urban transportation system is an intermingled system with many variables and loops (feedback) between subsystems, and affects the topology of transportation. The conventional linear quantitative technique should not be used to characterize the properties of this complicated system. As a result, in this work, the system dynamics (SD) technique is employed to mimic the development of the urban transportation system.

Interactions between numerous such loops govern the dynamics of most systems. It can be demonstrated that different sets of such loops (and delays) lead to distinctive patterns of underlying dynamical activity (e.g., exponential growth, goal-seeking, and oscillation). The analysis and understanding of the behavior of systems regulated by such loops (and delays) are referred to as system dynamics, and there is a highly well-developed and large academic literature on the methodology and modeling approaches used for this purpose as shown in Figure 3.1.

Forrester from MIT created the SD in the 1950s. The method is employed in quantitative studies of the complex socioeconomics sector and is defined by the feedback control theory, coupled with computer simulation technology. Equations, variables, and feedback loops are used to implement SD techniques. The variables are:

- level variable (defines a flow over continuous periods);
- rate variable (defines a flow over a specific period);
- auxiliary variable (it identifies the rate variables).

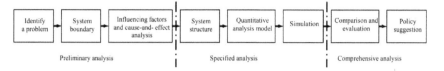

Figure 3.1   System dynamic approaches.

Equations in integral, differential, or other forms connect the three main types of variables. The complexity and breadth of the urban transportation system make it unsuitable for simulating and studying it using conventional methods. Therefore, complex system analysis has used the SD technique. The first SD model was put out by Forrester and is used to model urban systems. The SD technique is often used to assess the effectiveness of local, sustainable development, examine the connection between land use and transportation, and calculate the environmental effects of industrial gardens.

Three steps may be identified when adopting an SD approach: preliminary, defined, and thorough analysis. In the early analysis, it is vital to specify the external and internal variables, particularly the casual feedback loops of the variables, as the knowledge of the system features deepens. Based on the findings of the preliminary analysis, the system structure is built in the defined analysis, and coefficients and equations are provided to carry out a quantitative simulation process.

3.2 Feedback Loops

Input data influence some elements, which influence some other components, and so on until the elements that compute the intended outputs of the model are reached; this is a causal chain, and it is observable in simple models. In terms of pictorial representation:

However, more complex structures have components for whom the response may influence one of their inputs in some way. As a result, we have a system with a circular layout, such as:

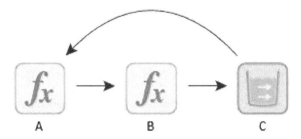

The above pictorial representation is termed a feedback loop. As seen above, the effects of one variable (A) have an impact on another variable (B), which in turn has an impact on a third variable (C), which in turn "feeds back" to influence A, and so on. What feedback loops show is a recursive series of events. It is important to keep in mind that the phrases "feedback" and "cause and effect" are used on purpose to indicate that the connection between the variables is dynamic and that the system develops over time (though we will learn in a later lesson that systems with feedback loops may also establish a dynamic equilibrium) [4].

Feedback loops may be either positive or negative.

Self-reinforcing, positive feedback loops are the best kind. When modifications are accompanied by growth, the positive feedback loop acts as an amplifier. If you have enough adults, they will have babies, and those babies will have babies, and so on, until the whole planet is populated with rabbits (or there are negative feedback loops that cancel out this good one).

When anything goes wrong, a negative feedback loop fixes it. Systems tend toward stability and balance via the action of negative feedback loops. There will be more rabbit deaths due to a lack of food if there are too many rabbits (hence, you have fewer rabbits).

Feedback structure; causing loop diagrams are the pictorial presentation of the systems used in the system dynamics approach [5]. This is how the causal loop diagram for the two loops above may look:

A "+" or "−" at the end of each arrow denotes a positive or negative causal relationship between the shown variables. If there is a positive correlation between two variables, then it means that more of one thing may be expected to happen as a result of changing the other thing (and, vice versa, as the cause decreases, the effect decreases) [6]. There is a negative correlation between these two variables since a rise in one will hurt the other (and, vice versa,

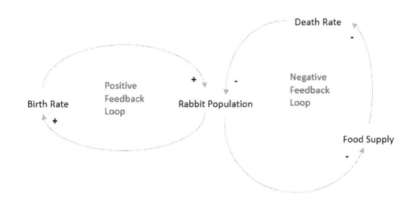

Figure 3.2   Causal loop diagram example.

as the cause decreases, the effect increases) [6]. It has been shown in the Figure 3.2 that a loop is a positive feedback loop if and only if the number of negative connections in the loop is even, and that it is a negative feedback loop otherwise.

As a result, we can see that the left-hand loop is a positive feedback loop in this basic system: a larger rabbit population leads to a higher birth rate, which in turn leads to a larger rabbit population. It is a growth factor. If there are more rabbits, the food supply goes down; if there are fewer rabbits, the mortality rate goes up; and if the death rate goes up, the rabbit population goes down. This serves as a check on the rate of population increase. There is a mutual influence between these two loops, and their combined effects may take on a variety of forms depending on the values of the rate parameters [7].

The system has been oversimplified, of course. To make it more lifelike, we may include more loops. Predators, for instance, could have an influence on the mortality rate of rabbits (and the birth and mortality rates of predators may be affected by the fluctuating rabbit population). It is also important to remember that certain loops will likely have delays. For instance, the mortality rate may not instantly rise if food supplies were cut. It could be a while before we see results from this. The system's dynamic complexity would increase as a result of the addition of these elements [8] as shown in the Figure 3.3.

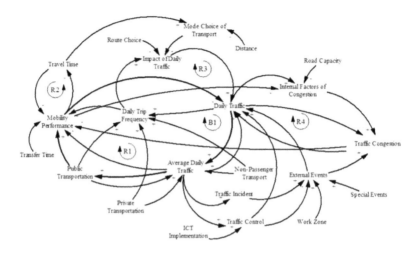

Figure 3.3   Daily traffic volume, overcrowding, and transportation implications: causal loop diagram [9].

3.3 Urban Transportation System (UTS) Flowchart

The following is a flowchart of the urban transportation system, as shown in the Figure 3.4 based on the aforementioned study and taking into account the defining features of such a system.

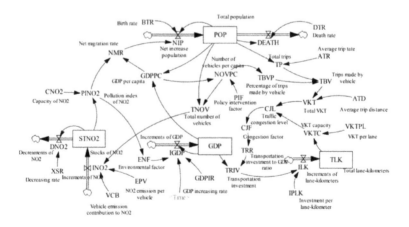

Figure 3.4   Urban transportation system [3].

3.4 Population Sub-model

It is the goal of this model's population sub-model to accurately portray the present state of urban development in each city. The size and composition of metropolitan populations have significant effects on public transportation systems that serve their residents [9].

The population density of a city is directly proportional to the number of people who live there [10]. The reason for this is obvious: people are what keep metropolitan life ticking around. A growing population and a thriving economy are two of the many variables that might influence the overall demand for transportation. Yet environmental factors, especially air quality, have a major role in shaping human movement patterns.

The number of people in the world has been designated as the level variable, with the rates of birth, death, and migration into and out of the country functioning as independent variables. After deciding to use the whole population as a level variable, this choice was made. Both the GDP per person and the quality of the environment as a whole affect the population's tendency to move around. Correlation analysis may help us figure out how each of these factors contributes to the whole [11].

The respective equation for the model is:

$$L\ POP.K=POP.J+KT^*(NIP.JK-DEATH.JK) \qquad eq\ (1)$$

$$R\ NIP.JK=POP.J^*(NMR.J+BTR) \qquad eq\ (2)$$

$$R\ DEATH.JK=POP.J^*DTR \qquad eq\ (3)$$

$$\underline{A}\ NMR.J=f\ (GDPPC.J,\ PINO2.J) \qquad eq\ (4)$$

3.5 Economy Sub-model

Growth in the number of persons purchasing their vehicles is directly related to the tremendous economic expansion that has occurred over the last several years. When contrasted during the same period, the number of additional cars sold in emerging countries in Asia is much higher than the number of cars sold in the region's industrialized nations. As a consequence of what

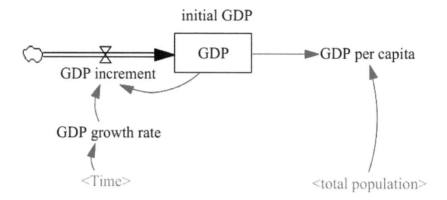

Figure 3.5   Economy sub-model diagram.

happened, there has been a shift away from the usage of non-motorized vehicles and toward the use of motorized vehicles.

This result also shows in the Figure 3.5 how the transportation system may be significantly impacted by the economy's fast growth, particularly the acceleration of the rise in GDP per capita. The fact that this result supports both of these claims serves as evidence for this.

The main factors influencing the growth of the urban transportation industry are described in the economics sub-model. The level of economic growth, which directly affects the number of people moving both into and out of the city, is one measure of a city's competitiveness. This flow of individuals demonstrates the city's capacity to draw in and keep citizens. It encourages the purchase and use of personal automobiles, which ultimately results in a rise in the total number of vehicles on the road [12].

The economy is linked to transportation investment because, to maintain a stable traffic situation, the government must expand its financial investment in transportation infrastructure. This is because, to maintain a stable traffic condition, the government must enhance the financial input it has. The level variable will be the GDP, the rate variable will be the GDP's annual increase, and the auxiliary variables will be the GDP's annual growth rate, the rate of transportation investment, and the environmental component [13].

The GDP will serve as both the level and the rate variable, and its annual growth rate will serve as the rate variable. The environmental influencing index is a statistic that assesses how much changes in environmental quality have an impact on GDP growth. The following are the individual equations that make up this sub-model:

$$L\ GDP.K=GDP.J+IGDP.JK \qquad eq\ (5)$$

$$R\ IGDP.JK=GDP.J*GDPIR.J*ENF.J \quad eq\ (6)$$

$$A\ TRIV.K=GDP.K*TRR.K \qquad eq\ (7)$$

3.6 Vehicle Ownership Model

Due to its centrality and interconnectedness with the rest of the proposed model's constituent parts, the sub-model displaying the total number of cars is at the heart of this investigation. In addition, all of the supplementary models provide the total number of cars. More than that, it is this particular sub-model that links together the others. The expansion of the economy and the subsequent rise in transportation needs may be linked to the rise in the number of automobiles on the road, and this is a possibility with a high degree of plausibility.

Given the high degree of probability linking these two phenomena, it seems reasonable to conclude that this is the case; this conclusion is further supported by the aforementioned fact. However, as the economy and the human population grow, a rise in the usage of vehicles will cause traffic congestion and an increase in air pollution. The expansion of both the economy and the human population is directly responsible for this. We have settled on using the overall number of vehicles, the number of cars per capita, and the component of policy action as our supplementary variables. The equations that make up the various components of this sub-model are as follows:

$$A\ TNOV.K=GDP.J+IGDP.JK\ eq(8)$$

$$A\ GDPPC.K=GDP.K/POP.K\ eq(9)$$

The number of automobiles per capita and the per-capita gross domestic output are significantly correlated. This association is substantial and favorable. The quantitative relationship between the average number of automobiles per person and each person's gross domestic product may be shown using the following equation:

$$A\ NOVPC.K=\hat{U}*EXP(\hat{u}*EXP(\hat{I}*GDPPC.K))\ (10)$$

It is possible to determine the values of the coefficients by doing a correlation analysis on the data that is currently available on urban statistics. This will allow for the determination of the values. It is possible to perform this to calculate the coefficients of the equation. It is of the highest significance to keep in mind that the component of policy intervention has a direct effect on the number of vehicles owned by each person. This is because it has a direct influence on the number of automobiles owned by each individual.

As a result of this, it is of the utmost significance to have this in mind. Because of this, the component that represents the policy intervention will be assigned the duty of acting as the control variable in the inquiry. This obligation will fall within the purview of the investigator. Because of this, we will have the chance to analyze the link that exists between the intervention policies and the ownership and usage of vehicles. The following ought to be the shape that the component of policy action should take that is going to be included in eqn (10):

$$A\ NOVPC.K=PIF*\hat{U}*EXP(\hat{u}*EXP(\hat{I}*GDPPC.K))\ (11)$$

3.7 Environmental Influence Sub-model

The environmental standards serve as roadblocks in the way of the growth of urban transportation, which stops it from progressing to more sophisticated levels. This is done to satisfy the prerequisites imposed by the natural world. NOx (occasionally the phrase NO2 is used for equivalent), CO, CH, and inhalable particulate matter are the most frequent types of pollutants that are emitted into the air when people drive their automobiles. In certain cases, the name NO2 is also used. Cyanide and carbon monoxide are two more major forms of environmental contamination (IPM).

The findings of the most recent study indicate that automobiles are responsible for the emission of more than 50% of the NOx that is produced in the major cities of China. As a direct result of this, one approach to calculating the degree of air pollution is to measure the amount of NO2 that is released into the atmosphere.

The amount of NO2 in stock was utilized as the variable for the level. The annual rise and decrease in NO2 concentration were the rate variables that were employed. The pace at which NO2 concentrations are falling, NO2 capacity, the NO2 pollution index, NO2 annual emissions per vehicle, and the environmental factor were selected to serve as auxiliary variables. The level variable that will be used is going to be the annual growth and decrease of NO2. The following equation is consider this as Equation.

$$L\ STNO2.K=STNO2.J+INO2.JK-DNO2.JK\ (12)$$

$$R\ INO2.JK=TNOV.J*EPV.J*VCB.J\ (13)$$

$$R\ DNO2.JK=STNO2.J*XSR\ (14)$$

$$A\ PINO2.K=STNO2.K/CNO2\ (15)$$

3.8 Transportation Sub-model

For this model, let us take the example of the city Beijing as shown in Figure 3.6. Only in the past decade has the population of Beijing expanded at an incredible pace. On the other hand, the overall number of people traveling has remained relatively stable at roughly 19 billion during the previous several years. However, as the population grew, the average number of trips taken by each person in a given year reduced by roughly 3.24%. The vast majority of people living in cities have converted from walking to motor forms of transportation, such as using public transportation or owning a vehicle. During the same period, the proportion of individuals who walked or biked decreased from 52.6% in 2002 to 26.5% in 2010, while the number of people who drove increased from 47.4% to 73.5%.

Buses provide a significant section of Beijing's vast public transportation system, making them an integral component of the entire system (BBS, 2012). In 2002, its networks accounted for 82.7% of all public transit journeys. Despite rapid expansion in rail transit in the city, bus travel remained a significant component of Beijing's total public transportation usage in 2010. The system's ability to handle the ever-increasing demand for bus journeys has come under heightened criticism in recent years. We surveyed a sample population utilizing a questionnaire given through an online platform (http://www.sojump.com/) to better understand how people in Beijing's metropolitan areas perceive the city's buses and to discover flaws and weaknesses.

Our objectives were to get a better understanding of how inhabitants of Beijing's urban areas view the city's buses, as well as to identify flaws and weaknesses. The poll had 214 replies, all of which were completed. Based on the survey responses, we were able to discover the five most essential characteristics of bus service that passengers are concerned about timeliness (0.268), maximum waiting time (0.266), average speed (0.22053), price (0.0187), and other considerations (0.074) [14]. According to the survey findings, even those who own vehicles may consider switching to public transportation, given that bus service has grown by 76.7%. The survey findings indicated that this is the case.

Appendix A offers a list of specific equations that may be solved. The yearly trip volumes per person for the years 2002−2010 were derived from the China Statistical Yearbook, as were the data for the number of automobiles in use, bus trips made rail transit length, rail transit users, and rail transit vehicles.

The taxi ride data were obtained from the same source and for the same period. Furthermore, the statistics for these variables were gathered by consulting the China Statistical Yearbook. Operating costs accounted for just 0.8% of overall expenditures, according to our best estimates. Based on surveys and studies conducted by the Beijing public transportation administration, the following data was gathered:

The average time spent running was 40 minutes, the average number of runs was 20, and the time interval was an hour. The fee for equipment updates was 3.5%. The cost of one bus unit was 500,000 RMB. The scrapping rate for buses was 1%. The cost of maintaining the bus was 5% of the overall cost. Based on the questionnaire data and our calculations, we determined that the bus had an appealing value of 1.304, was punctual 88% of the time, and traveled at an average speed of 28 km/h, the other features accounted for 70% of the total, and that its appeal factor was 1.28.

A dynamic equilibrium is maintained between the supply side of the transportation demand equation and the demand side thanks to the amount of urban infrastructure development, which is a reflection of the level of available transportation options.

The quantity of new urban infrastructure construction is sufficient to preserve the dynamic equilibrium. It has been shown that the amount of money invested in the building and improvement of infrastructure is directly related to the number of different kinds of transportation that may be utilized concurrently. The total lane length will serve as the level variable, the annual increment of the lane length will serve as the rate variable, and the auxiliary variables will consist of transportation investment as a percentage of GDP, investment per lane, vehicle-kilometer-traveled capacity per lane, and overall vehicle-kilometer-traveled capacity.

The total length of the lane will serve as the level variable, while the yearly percentage increase in the length of the lane will serve as the rate variable. The equations for following system are:

A TRIV.K=GDP.K*TRR.K (19)
L TLK.K=TLK.J+ILK.JK (20)
R ILK.JK=TRIV.J*IPLK (21)
A VKTC.K=TLK.K*VKTPL (22)

3.9 Environmental Sub-model

The environmental rules place restrictions on the expansion of urban transportation and prohibit it from increasing further and more rapidly. When people drive their automobiles, the most common types of air pollutants that are released into the atmosphere are nitrogen oxides (NOx), carbon monoxide (CO), and carbon monoxide (CH) together with inhalable particulate matter (NOx; the designation NO2 is frequently used for similar) (IPM) [15].

According to the most current research, it is assumed that vehicles are responsible for emitting more than 50% of the NOx that is created in China's largest cities. As a consequence of this, the measurement of NO2 emissions may serve as a method for estimating the level of air pollution that is currently present. As auxiliary variables, we chose the following: the rate of NO2 reduction, the NO2 capacity, the NO2 pollution index, the yearly NO2 emissions per car, and the environmental factor. It was decided that the yearly increase and drop of NO2 would serve as the level variable, while the annual increment and decrease of NO2 would serve as the rate variables. Equations such as the ones listed below make up this component of the model:

L STNO2.K=STNO2.J+INO2.JK–DNO2.JK (12)
R INO2.JK=TNOV.J*EPV.J*VCB.J (13)
R DNO2.JK=STNO2.J*XSR (14)
A PINO2.K=STNO2.K/CNO2 (15)

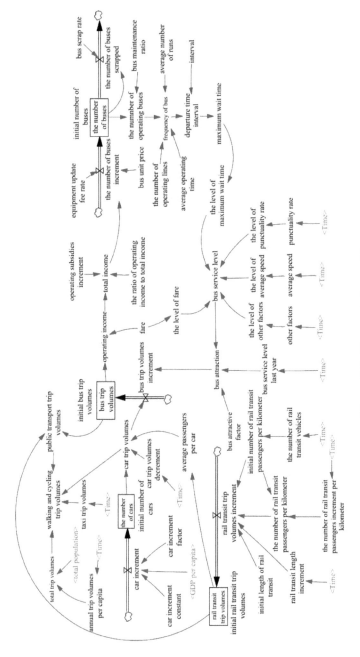

Figure 3.6   Transportation model.

When determining the NO2 pollution index, the environmental component is one of the considerations that are taken into account. One method, known as correlation analysis, is used to ascertain the shape that the index will take. This kind of indication is used to assess the current level of pollution that can be found.

3.10 Case Study

Researchers deploy Vensim PLE and feed its data to Dalian Central City to test the SD model created for the project. A significant city in the Liaoning region is Dalian, the provincial capital of Liaoning in eastern China.

Besides serving a crucial function as the province's administrative hub, Dalian Central City has a population of about 1.92 million people and covers an area of 248 square kilometers. According to data gathered in 2004, the great majority of visitors to Dalian City commute using public transportation. Public transportation merits some consideration because it accounted for more than half (40.96%) of all modes of transportation. To fully understand how the policy of automobile ownership has impacted the development of urban transportation infrastructure over the course of the motorization process. This is true since the raise of the automobile is a sign of the motorization of society.

3.10.1 Modeling factors

Given that the proposed model is highly complex and makes use of an excessive number of variables, it is necessary to determine their values as well as the coefficients that are linked with them. Estimating variables and coefficients can be done in one of three ways, depending on the specific situation: To accomplish this goal, you can (i) make use of the findings of statistical research, (ii) examine reports that have already been made public, or (iii) make use of correlation analysis by way of the application of mathematical and statistical models.

The calculated coefficients for the net migration rate (NMT) and the number of vehicles per capita (NOVPC) are presented in Table 3.1. In Table 3.1, a selection of feasible coefficient values from the vast pool of options is presented for usage.

Table 3.1   Values of Parameters

Parameters	Value	Unit
Birth rate	0.004	-
Death rate	0.006	-
GDP increasing rate	0.12	-
Average trip rate	2.1	-
Average trip distance	4.55	Kilometer
VKT per lane	800	VKT
Investment per lane-kilometer	10	Million Yuan RMB
NO_x emission per vehicle	20	Kg/ vehicle/year
Vehicle emission contribution to NO_x	0.5	-
NO_x decreasing rate	0.2	-
NO_x capacity	75.6	Thousand tons

Table 3.2   Comparison of model outputs with reported data

Index	Model output	Reported data	Error (%)
Population (million)	1.91669	1.9195	−0.15%
GDP (billion yuan)	34.1735	35.887	−4.77%
Number of vehicles	11,1402	11,6650	−4.50%

3.10.2 Model simulation inspection

The output in 2005 is replicated by using time series data spanning the years 2000−2004, to determine how accurately the suggested model represents reality. This is done to examine the level of accuracy that the model can provide and to put it through its paces.

This helps to validate the model more comprehensively. Because the validation only uses data from a single instant in time, the year 2005, it provides the appearance that it is not based on scientific standards. On the other hand, the fact that this is the case is that it is difficult to locate information that is both trustworthy and beneficial. In Table 3.2, a comparison is made between the output numbers for population, GDP, and the number of cars, and the figures that were reported.

The results of this comparison are shown in Table 3.2.

According to the figures, less than 5% of the population, GDP, and total number of vehicles are inaccurate. It has a great degree of accuracy and can imitate the suggested model.

3.10.3 System simulation

As was previously said, the main goal of this line of research was to look at how government policies affected the development of metropolitan regions and the transportation networks connected to such regions. The component that decides whether or not the policy will intervene has been determined to be the control variable that will be employed in the simulation of the proposed model (PIF).

The current model, which has been in use for quite some time, will have a lifespan extension of 50 years starting in the year 2000. The time step has been set at one year to make it easier to grasp the simulation's outcomes. The optimum way to boost the manufacture of automobiles has been determined after an analysis of five possible policy scenarios. The options include strongly encouraging, encouraging, severely limiting, restricting, or not participating at all.

For the first three and the last three, respectively, the relative values of the PIF control variables that are related to them are 1.0, 0.8, 0.5, and 1.5. Figures 3.7–3.10, which include information on the environment, population, gross domestic product, and the number of autos, display the findings.

As can be seen in Figure 3.10, the overall number of automobiles continues to increase in the shape of an "S curve" throughout the duration of the period in question. Despite this, the actions of the government have had a significant impact on the expansion of the car sector. The characteristic outcome is that the curve vacillates less often when the emphasis of a government action moves from encouragement to restriction. This is because restricting behavior has the opposite effect of encouraging behavior. For example, the number of cars sold skyrockets in the first 20 years under the

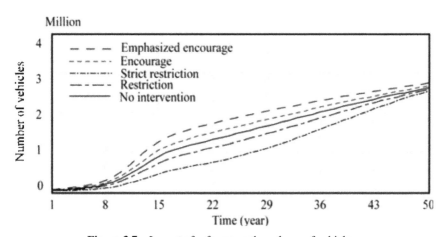

Figure 3.7 Impact of reforms on the volume of vehicles.

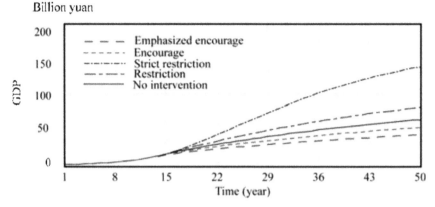

Figure 3.8   Impact of reforms on GDP.

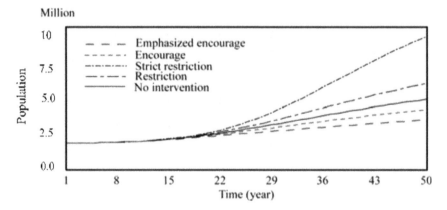

Figure 3.9   Impact of reforms on population.

stressed-encouraging policy before gradually decelerating in the following 30 years under the severe limiting policy. However, under the stressed-encouraging policy, the number of cars sold initially grows slowly before picking up speed in the following 30 years. There is no discernible difference in how either approach would affect the total quantity of autos produced.

The effect that government initiatives have had on GDP is beginning to become clearer as more time passes. When the tight limiting regime is in place, the rate at which the GDP grows is far quicker than it would be under any of the other alternative policy scenarios. Improving the level of service that is offered by public transportation is the most effective tactic for reducing

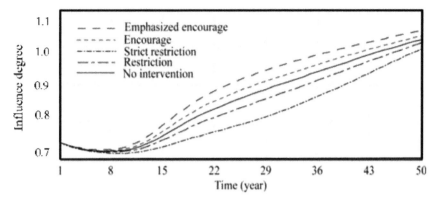

Figure 3.10 Impact of reforms on the environment.

the number of people who own automobiles and the amount of time they spend driving those vehicles. When you consider that public transportation accounts for 50% of all journeys taken in Dalian, there may not be much room left.

The policy has an effect on the population in a manner that is analogous to that of GDP's influence. Because of the stringent regulations that are placed on cars, an increasing number of people are opting to make use of public transit as opposed to driving their autos. As a direct consequence of this, urban density is rapidly increasing, which raises the issue of whether or not this is beneficial for the expansion of metropolitan areas. There is a one-to-one relationship between emissions from automobiles and pollution in the atmosphere. Legislation must be enacted that places a limit on the number of vehicles that may be owned by a single individual. This will help to mitigate the negative impact that urbanization and pollution have on the natural world.

3.11 Summary

In the course of this analysis, a few possible future solutions that make use of intelligent mobility modeling to enhance the efficiency of transportation networks will be investigated and assessed. This investigation's objective is to explore possible solutions that may be used to enhance the effectiveness of the transportation infrastructure that is already in place. The findings of this investigation could provide researchers with the means to discover prospective strategies for enhancing the operational efficiency of transportation network systems. When intelligent mobility modeling is employed, a wide range of possible solutions to the issue of decreasing congestion while simultaneously boosting mobility may be tested and assessed. This allows for a greater scope of potential solutions to be considered. This is a result of the fact that it is practical to test and evaluate a wide variety of different alternative solutions. As a direct consequence of this, it is now feasible to investigate and assess a broad variety of prospective solutions. It is possible to reduce congestion by the implementation of a variety of different methods, such as the management of demand and capacity, the implementation of congestion pricing, and the use of information and communications technology for traffic and route controls. These are just some of the options available.

Utilizing a simulation model that is predicated on system dynamics to achieve the primary objective of this project, which is to contribute new information to the existing body of knowledge about the planning and management of urban transportation, is the key to achieving this objective. This will be accomplished via the provision of a solution that is based on system dynamics. One of the original contributions that may be made to the modeling of urban mobility is the identification of the factors that play a role in the chain of causality connecting the variables that are part of the CLD. One of the fascinating adjustments that may be done is the use of this particular capability. The creation of several potential future scenarios is yet another fresh idea that might be done. This study contributes to the corpus of research that has previously been conducted by investigating the connections discussed earlier in a setting that is both theoretical and practical via the development of models and potential outcomes. This study contributes to the body of prior research since it investigates the connections between concepts in both theoretical and practical settings. As a consequence of this, the initiative may be beneficial not just to the academic community but also to the general public.

References

[1] "Download Rodrigue J-P., Comtois C., Slack B. The Geography of Transport Systems [PDF] - Sciarium." https://sciarium.com/file/11 7738/(accessedJan.02,2023).

[2] R. Maršanić and L. Krpan, "Contemporary Issues of Urban Mobility," Int. J. Vallis Aurea, vol. 1, no. 2, pp. 5–14, Dec. 2015, doi:10.2507/IJ VA.1.2.1.12.

[3] J. F. Wang, H. P. Lu, and H. Peng, "System Dynamics Model of Urban Transportation System and Its Application," J. Transp. Syst. Eng. Inf.

Technol., vol. 8, no. 3, pp. 83–89, Jun. 2008, doi:10.1016/S1570-6672 (08)60027-6.

[4] "Lecture 13 Dynamic analysis of feedback Closed-loop, sensitivity, and loop transfer functions Stability of feedback systems".

[5] J. Miguel and M. Pópulo, "High-Tech Diagnostic Imaging Clinical Decision Support Tools Adoption: Study using a System Dynamics Approach At the Massachusetts Institute of Technology Engineering Systems Division," 2010.

[6] G. Tircsóet al., "2 Gadolinium(III)-Based Contrast Agents for Magnetic Resonance Imaging. A Re-Appraisal," Met. Ions Bio-Imaging Tech., pp. 39–70, Mar. 2021, doi:10.1515/9783110685701-008/HTML.

[7] "A behavioral approach to feedback loop dominance analysis - Ford - 1999 - System Dynamics Review - Wiley Online Library." https://onli nelibrary.wiley.com/doi/pdf/10.1002/%28SICI%291099-1727%2819 9921%2915%3A1%3C3%3A%3AAID-SDR159%3E3.0.CO%3B2-P (accessedJan.02,2023).

[8] D.-H. Kim, "Kim, Dong-Hwan, 'A New Approach for Finding Dominant Feedback Loops: Loop By Loop Simulation for Tracking Feedback Loop Gains.'" 1995. Accessed: Jan. 02, 2023. [Online]. Available: https: //archives.albany.edu/concern/daos/nk322z34c?locale=en

[9] E. Suryani, ;Hendrawan, ; Adipraja, R. Indraswari, A. Stmik, and I. Malang, "SYSTEM DYNAMICS SIMULATION MODEL FOR URBAN TRANSPORTATION PLANNING: A CASE STUDY," Int j simul Model, vol. 19, pp. 5–16, 2020, doi:10.2507/IJSIMM19-1-493.

[10] "Operations support and logistics," Eng. Syst. Acquis. Support, pp. 101–137, Jan. 2015, doi:10.1016/B978-0-85709-212-0.00007-9.

[11] S. R. Golroudbary and S. M. Zahraee, "System dynamics model for optimizing the recycling and collection of waste material in a closed-loop supply chain," Simul. Model. Pract. Theory, vol. 53, pp. 88–102, 2015, doi:10.1016/j.simpat.2015.02.001.

[12] S. V. Sharif, P. H. Moshfegh, M. A. Morshedi, and H. Kashani, "Modeling the impact of mitigation policies in a pandemic: A system dynamics approach," Int. J. Disaster Risk Reduct., vol. 82, Nov. 2022, doi:10.101 6/j.ijdrr.2022.103327.

[13] R. Rehan, M. A. Knight, A. J. A. Unger, and C. T. Haas, "Financially sustainable management strategies for urban wastewater collection infrastructure - development of a system dynamics model," Tunn. Undergr. Sp. Technol., vol. 39, pp. 116–129, Jan. 2014, doi:10.1016/j.tu st.2012.12.003.

[14] Y. Tan, L. Jiao, C. Shuai, and L. Shen, "A system dynamics model for simulating urban sustainability performance: A China case study," J. Clean. Prod., vol. 199, pp. 1107–1115, Oct. 2018, doi:10.1016/J.JC LEPRO.2018.07.154.

[15] S. V. Sharif, P. H. Moshfegh, M. A. Morshedi, and H. Kashani, "Modeling the impact of mitigation policies in a pandemic: A system dynamics approach," Int. J. Disaster Risk Reduct., vol. 82, p. 103327, Nov. 2022, doi:10.1016/J.IJDRR.2022.103327.

4

Deep Learning Methods for High-level Control using Object Tracking

P. Kanmani[1], N. Yuvaraj[2], S. Karthic[3], Sri Preethaa K. R.[2], and Min Dugki[4]

[1]Department of Data Science and Business Systems, SRM Institute of Science and Technology, Kattankulathur, Chengalpattu, India
[2]School of Computer Science and Engineering Vellore Institute of Technology, Vellore, India
[3]Department of Computer Science and Engineering, KPR Institute of Engineering and Technology, India
[4]Department of Computer Science and Engineering, Konkuk University, South Korea
E-mail: pkanmaniit@gmail.com; yuvaraj.n@vit.ac.in; karthic.s@kpriet.ac.in; sripreethaa.kr@vit.ac.in; dkmin@konkuk.ac.kr

Abstract

The utilization of unmanned aerial vehicles (UAVs) has witnessed a recent surge in popularity, particularly within the realms of computer vision (CV) and remote sensing. Notably, advancements in object identification and tracking techniques have found widespread application in various UAV-related endeavors, encompassing activities such as environmental monitoring, precision agriculture, and traffic management. Furthermore, UAVs have introduced innovative approaches to deep-learning-based threat identification and recognition, which have been scrutinized within the context of the military and defense sectors. Within the spectrum of deep learning techniques, including deep belief networks (DBNs), auto-encoders (AE), convolutional neural networks (CNNs), and others, this study embarks on a comprehensive analysis of the current state-of-the-art and future potential of deep-learning-based UAV object recognition and tracking methodologies. This chapter is anticipated to provide remote sensing researchers with an overview of DL-based UAV object recognition and tracking techniques while simultaneously igniting inspiration for their prospective advancements.

Keywords: Unmanned aerial vehicles, deep learning, object detection, object recognition

4.1 Introduction to Object Tracking

Object tracking is a deep learning process that involves estimating and forecasting the movement of an object within an environment. It relies on historical data related to the object's behavior to predict its future positions or goals. This tracking process begins by detecting the object's presence initially and then continually monitoring and updating its position. Notably, object tracking techniques can be extended from two-dimensional (2D) objects to encompass three-dimensional (3D) objects [1]. Regardless of the object's dimensions, the algorithm operates by identifying the object within an image or video stream and providing precise classification, along with highly accurate predictions of its future path.

In practice, the tracked target is typically enclosed within a square or rectangle, creating a well-defined bounding box that encompasses all of its edges. This visual representation aids users in easily comprehending the object's location and movement within the application [2]. Furthermore, the identified object is often accompanied by a label that describes its nature, whether it is a vehicle, an animal, or a person.

4.1.1 Levels of object tracking

There are many related directions for visual object tracking, such as single-object tracking [3], multi-object tracking [4], 3D object tracking [5], and video object segmentation [6]. Figure 4.1 shows the different types of object tracking.

4.1.1.1 Single object tracking

Single object tracking (SOT) is targeting one single object in an image. The attempt of finding the multiple targets is not needed in these types of images [7]. A sample of the image dataset is given in Figure 4.2. It follows a single target in an image or video sequence. Steps involved in SOT are as follows:

1. Extracting features from the input image

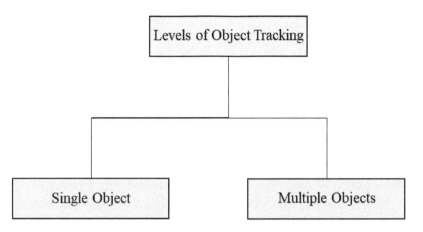

Figure 4.1 Levels of object tracking.

(a) PASCAL VOC   (b) ILSVRC   (c) Open image   d) COCO 2017

Figure 4.2 Sample image dataset.

2. Generating target areas
3. Building the model
4. Updating the model periodically

Challenges in single object tracking:

The following are some of the challenges in SOT:

1. Targeting a single object among complex objects is little complicated
2. Too many interferences
3. Target deformation
4. Similar objects
5. Scale transformation
6. Low quality resolution

Figure 4.3   Applications of multiple target tracking.

4.1.1.2 Multiple object tracking

Multiple objects tracking (MOT) otherwise called multiple target tracking is defined as the locating of more than one object and extracting features of each target object to classify the label for input video. The bounding box of an object is similar to single target object; the only difference is that the bounding boxes are overlapped as shown in Figure 4.3. The overlap is encouraged to identify more than one target object in a single shot [8]. The MOT also has few challenges, which are:

1. Frequent occlusions
2. Initialization and termination of tracks
3. Similar appearance
4. Variations due to geometric changes
5. Non-linear motion
6. Low quality resolution

4.2 Unmanned Aerial Vehicle

It is an electromechanical vehicle that does not need manual intervention to operate on board. Depending on pre-programmed programming, unmanned vehicles can operate autonomously or remotely. These advanced features help us to indulge UV in several areas of computer applications [9]. Different types of UAV are shown in Figure 4.4.

Unmanned aerial vehicles (UAV) are commonly called drones, which are considered to be the predominant type of unmanned systems (US). DRONE is abbreviated as "Dynamic Remotely Operated Navigation Equipment". An unmanned vehicle (UV) may fly across the air, examining vast regions and go to locations where people are hostile. Operation of UAV includes an automated path planning, obstacle avoidance, takeoff, and landing [10].

Applications of unmanned vehicles include:

1. Security, monitoring, and surveillance

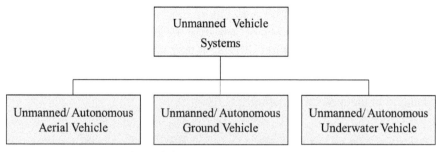

Figure 4.4 Types of unmanned vehicles.

2. Disaster management
3. Remote sensing
4. Search and rescue (SAR)
5. Construction and infrastructure inspection
6. Precision agriculture
7. Real-time monitoring of road traffic
8. UAVs for automated forest restoration
9. UAVs for inspection of overhead power lines

4.3 Types of Unmanned Aerial Vehicles

Drones can be categorized based on the device size and weight as shown in Figure 4.5. The size and weight of the drones are varied depending upon the applications they are used. UAVs can be as little as insects or as large as airplanes, and they can weigh a few hundred grams or several kilograms. Power supply in drones prefers battery power more frequently than fuel power. The majority of UAVs weigh around 30 kilos, which is said to be

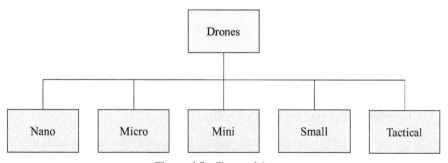

Figure 4.5 Types of drones.

the light-weight UAV [11]. This uses multirotor aerodynamic shapes than fixed wings, and has flying times of less than an hour on average. Radio waves are used to remotely control drones, or they can fly themselves along a predetermined course [12]. Small and light-weight drones are preferred as the main type for the monitoring and controlling application and they have high demand in the industrial market.

4.4 Classification of UAVs

UASs do not have any common standard and so different sectors have framed their own set of standards. The military sector has employed a tier system based on which people classify UAS according to size, range, altitude, and rotors, which is given in Figure 4.6.

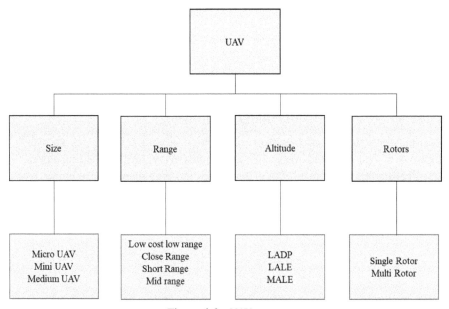

Figure 4.6   UAV types.

4.4.1 Classification based on the range

UAVs are classified into four major categories based upon their size such as micro-UAVs, small UAVs, medium UAVs, and large UAVs.

Figure 4.7 Nano drones.

4.4.1.1 Micro-UAVs

In this class, the size of the UAVs ranges from 0.6 grams to 30−50 cm long. The mass of the drone is almost equal to an insect and a large insect in some cases with fluttering or rotary wings [13]. Perching and landing on small surfaces is empowered by the fluttering design of the drone while the rotary wings enable the drones to fly in narrow spaces among obstacles.

The minuscule size of the drones and its weight is more advantageous in spying and biological warfare applications and also some larger UAV uses in aircraft configurations. Examples of very small UAVs include the DJI Mini, Skeyetech, and Kespry 2S, as depicted in Figure 4.7 [14, 15].

The demand for micro- and nano-UAVs is expected to peak by the year 2030, a need that gained prominence during conflicts such as the Afghanistan and Gulf wars. The future of warfare is envisioned to involve battles employing electronic countermeasures, space-based craft, and UAVs. A notable illustration of micro-UAVs is the renowned Black Hornet Nano, developed by Prox Dynamics AS of Norway. This miniature UAV is currently in use by armed forces in various countries, including the United States, the United Kingdom, France, and Germany.

4.4.1.2 Small UAVs

UAVs with a size range between 50 cm and 2 m are classified as small UAVs. Most of the drones in this category are based on the fixed-winged

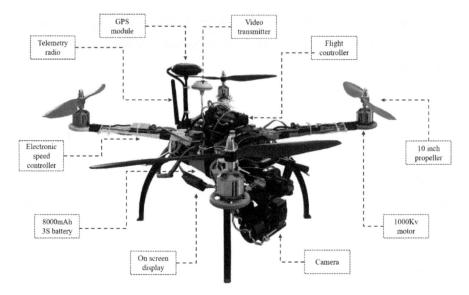

Figure 4.8   Small UAV.

hand-launched model by flinging it in the air as shown in Figure 4.8 [8]. The RQ-11 Raven with 1.4 wingspan designed by US, The Turkish Bayraktar with a data link range of 20 km and weighing approximately 5 kg, and the RQ-7 by US Army and RS-16 by American Aerospace are all some of the examples of small-sized UAVs.

4.4.1.3 Medium UAVs

The drones of this category weigh lesser than an aircraft but make it highly difficult to be carried by one person. They usually have a wingspan of about 5–10 m and can carry payloads of 100–200 kg. The Israeli-US Hunter, the UK Watchkeeper, RQ-2 Pioneer, the BAE systems Skyeye R4E, and the RQ-5A Hunter are some of the examples of drones belonging to this category as shown in Figure 4.9 [9].

The Hunter has a wingspan of 10.2 m and 6.9 m and weighs around 885 kg at takeoff. The RS-20 by American Aerospace is another example of a crossover UAV that spans the specifications of a small- and medium-sized UAV. Medium UAV also has its applications in reading and scientific research assignments.

Figure 4.9 RQ-5A.

4.4.1.4 Large UAVs

The larger drones are certainly better than small drones owing to its higher autonomy, payload capacity, and their ability to go to a maximum altitude. Their size ranges between 20 and 61 ft long with a flight time ranging between 30 minutes to several hours. Larger drones have its applications in forest fire monitoring such as aerial photography, entertainment, science and several other assignments as shown in Figure 4.10 [16]. The UAV class applies to the large UAVs used mainly for combat operations by the military. Examples of these large UAVs includes the US General Atomics Predator A and B and the US Northrop Grumman Global Hawk.

Figure 4.10 General atomics predator.

4.4.2 Classification based on the range

The traveling ability of a drone is called the range, which is dependent on few parameters like amount of current produced by the aircraft, flight speed, and

Table 4.1   Category of UAVs.

Category	Weight (in kg)	Range (in m)	Altitude (in m)	Endurance (in h)
Close range (CR)	25–150	10–30	3000	2–4
Short range (SR)	50–250	30–70	3000	3–6
Medium range (MR)	Up to 1250	70–200	5000	6–10
Medium range endurance (MRE)	Up to 1250	>500	8000	10–18

endurance. The range is estimated using the following formula:

$$R = \frac{kV.\ V.\ 60.\ P}{12.5260}.E,$$
(4.1)

where R is the range, kV is the number of revolutions per minute, P is the propeller's pitch, and E is the endurance. Endurance is yet another critical factor defined as the total time of flight, which is highly dependent on size and weight of the aircraft. The endurance can be given by the following formula:

$$E = \frac{\text{Battery capacity}}{\text{Current}},$$
(4.2)

where E is the endurance of the aircraft. Based on the above factors, the US Military has classified the UAVs into five major categories such as

- Very close-range UAVs
- Close-range UAVs
- Short-range UAVs
- Mid-range UAVs
- Endurance UAVs

The specification of these categories is given in Table 4.1.

4.4.3 Classification based on altitude

Altitude refers to the height at which drones can fly. Based on the altitude, the drones are further classified into low-altitude platforms (LAPs), medium-altitude platforms (MAPs), and high-altitude platforms (HAPs). LAPs are usually economical and facilitate fast deployment due to which its applications are most likely in cellular communications. MALE UAVs with a flying altitude window between 10,000 and 30,000 ft are extremely useful in military or civil applications. Deployment of HAPs is much more complicated when compared to that of LAPs, and hence it acts as a tool to support

Table 4.2 The specifications of the three categories.

Category	Endurance	Flight	Range	Mass
Low-altitude deep penetration (LADP)	0.5−1	50−9000	>250	250−2500
Low-altitude long endurance (LALE)	>24	3000	>500	15−25
Medium-altitude long endurance (MALE)	24−48	3000	>500	1000−1500
High-altitude long endurance (HALE)	24−48	20,000	>2000	2500−5000

internet connectivity and can be deployed in critical situations like disaster relief activities [17]. The specifications of the three categories are given in Table 4.2.

4.4.4 Classification based on rotors

UAVs are also classified according to different wing and engine structures such as fixed-wing UAVs, rotary-wing UAVs, and flapping wing UAVs. The diagrammatic representation of the classification is given in Figure 4.11.

Fixed-wing UAVs are similar to migratory birds. Drones of this type stand vertically and fly at close arm length due to their reduced drift design. These

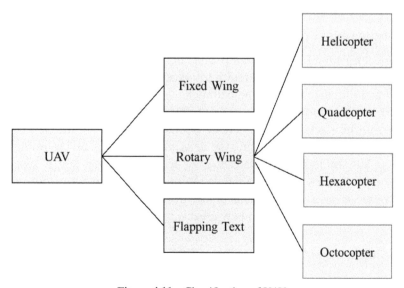

Figure 4.11 Classification of UAV.

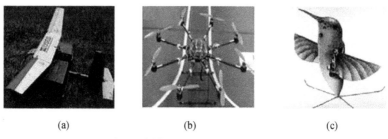

(a)   (b)   (c)

Figure 4.12   Classification of rotors.

drones have their applications exorbitantly in research and development. Rotary-wing UAVs use the force, enforced by the rotors to enable a balanced flight. The classification of drones is given in Figure 4.12.

These robots have their applications especially in robotic research and universities as they do not need any runways. Tricopters, quadcopters, pentacopters, hexacopters, and octacopters are various types of rotary-wing UAVs named according to the number of engines. These drones are operated manually and have a complex flight dynamic than any other type of drones. Flapping drones have a wide range of applications with vertical takeoff and hovering.

4.5 Object Detection from UAV-borne Video

4.5.1 Single object tracking

Single object tracking (SOT) is widely used in applications where the target is just one single object among multiple objects. The object of interest is identified and initialized before the commencement of the process. As soon as the first frame of the video is converted, this tracking technique gives the first frame to the tracker for it to create bounding box for the frame [18]. The tracker then focuses on locating the determined object in all the frames, which is a VISDRONE 2022 Dataset [19]. It is a detection-free tracking technique that requires manual initialization of constant number of objects. The technique is also called visual object tracking (VOT), which has the drawback of inability to deal with new objects appearing in the frames.

4.5.2 Multiple objects tracking from UAV-borne video

The task of identifying and tracking multiple objects within a video is known as multiple objects tracking (MOT) [20]. MOT involves not only detecting these objects but also representing their movements as trajectories with a high degree of precision [4]. This capability lays the foundation for various computer vision tasks, including video understanding, behavior recognition, and behavior analysis [21]. MOT has witnessed widespread adoption, thanks to the significant advancements in computer vision, and is now prominently utilized in domains such as human–computer interaction (HCI), intelligent video surveillance, and autonomous driving [5]. In many MOT applications,

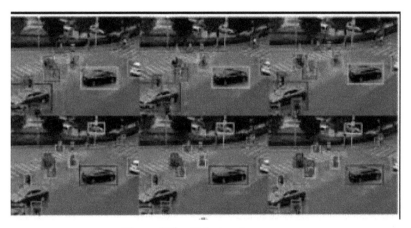

Figure 4.13 Object on the screen.

Figure 4.14 MOT is deployed with UAV.

there is typically a visual indicator around the object being tracked, such as a surrounding square or bounding box that dynamically follows the object's movements. This indicator serves to inform the user of the object's location on the screen, as illustrated in Figure 4.13.

Meanwhile, UAV has also seen a tremendous growth and owing to its flexibility and accessibility, MOT is deployed with UAV, and one such application is shown in Figure 4.14.

Various types of object tracking techniques have evolved into practice, which is discussed in the subsequent sections.

4.6 Tracking by Detection

The prevailing approach in object tracking, known as "tracking by detection" (TBD), involves detecting objects in consecutive frames and subsequently correlating them for re-identification. However, this method faces significant inefficiencies due to resource limitations, dynamic background variations, and the presence of densely packed small objects [22]. One notable application of TBD is in video tracking, where it plays a pivotal role in identifying moving objects within source videos. This capability is particularly valuable for processing live, real-time footage in domains such as security, military operations, transportation, and various industrial applications.

When deployed alongside UAVs, this technique involves capturing multiple images at regular intervals using onboard cameras. Ensuring substantial overlap between these images is crucial to prevent any omission of objects. This comprehensive process is referred to as photogrammetry, and it relies on metadata for image correlation, which is typically handled by a micro-computer on the UAV. Subsequently, structure from motion (SfM) software is employed to correlate or stitch together adjacent images, based on various parameters like angle matching and measurement [23]. The correlated images can then be harnessed for data analysis in various specific applications.

4.6.1 Multiple objects tracking based on memory networks

Video sequence has strong long-term correlation, and hence introducing a memory network to learn the timing information would be an ideal option. Thus, the deep neural networks such as recurrent neural network (RNN) and long-short term memory (LSTM) have stepped into the process of identifying multiple objects from videos [24]. The entire process of tracking of objects for a UAV captured drone is represented diagrammatically in Figure 4.15; the fusion of multi-frame features to improve the accuracy of video object detection is done by a new cross-framework of ConvLSTM.

Brain-inspired memory network, spatio-temporal memory network, and motion-aware network are some of the DL-based algorithms to track multiple objects in a UAV video input [25]. But all these techniques predict the trajectories in the next frame, whereas we need to update the outputs of the current frames for video object detection as in Figure 4.16.

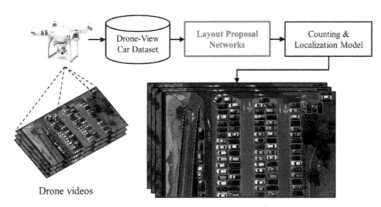

Figure 4.15 Tracking of objects for a UAV captured drone.

Figure 4.16 Video object detection (VID).

Siamese RPN and ReID output the current frames, making it feasible to process UAV videos.

4.6.2 Multiple objects tracking based on other techniques

Apart from the above-mentioned frameworks, techniques such as generalized graph differences (GGDs) [26] and context-aware IoU-guided tracker (COMET) [27] is also in practice for multi-object detection in videos. GGD uses a general tracker implementation especially for videos captured by UAVs, which is trained on VisDrone dataset. Even in an online video processing, GGD guarantees a global optimal solution. Context-aware IoU-guided tracker (COMET) uses multitask two-stream network and an offline reference proposal generation strategy for detecting multiple objects in the successive frames of the video footages. The technique generalizes the network

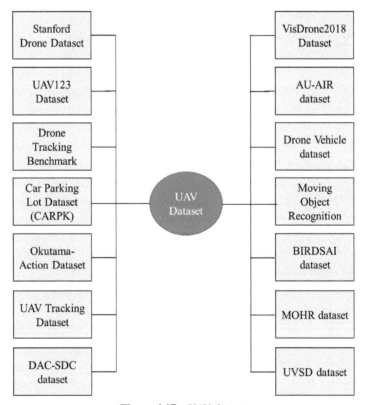

Figure 4.17   UAV dataset.

extremely well without adding any complexity to the model during online tracking. The method is proved to outperform state-of-the-art techniques in tracking small objects in any UAV captured videos.

4.7 UAV-based Benchmark Dataset

4.7.1 UAV datasets

With the data driven advancements many research works are proposed to build different datasets on various field for object detection and recognition that includes both images and videos as given in Figure 4.17. It also possesses, in this section, commonly used UAV-based remote sensing dataset for object detection and tracking is reviewed as given in Figure 4.18.

4.8 Deep Learning and UAV

Due to the technological development of deep neural network techniques, deep and complicated networks are capable of learning hierarchical feature representations given with sufficient sample data [28]. Deep neural network is said to be the conventional foundation for object tracking and detection based on UAV images as given in Figure 4.19.

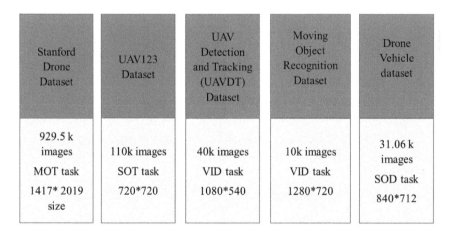

Stanford Drone Dataset	UAV123 Dataset	UAV Detection and Tracking (UAVDT) Dataset	Moving Object Recognition Dataset	Drone Vehicle dataset
929.5 k images	110k images	40k images	10k images	31.06 k images
MOT task	SOT task	VID task	VID task	SOD task
1417* 2019 size	720*720	1080*540	1280*720	840*712

Figure 4.18   UAV dataset with specification.

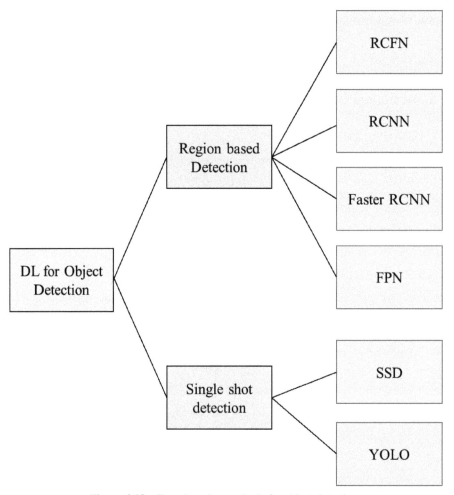

Figure 4.19   Deep learning methods for object detection.

4.8.1 UAV data based object detection and tracking using deeply supervised object detector (DSOD) Long short term memory

DSOD based object detection and LSTM object tracking is proposed [9].

(i) Object detection from the video taken by drone via RNN (recurrent neural network):

The two-stage object detector uses a selective search method to look for latent object locations (area proposals) in the image. In this method, the features were extracted independently by the network for each input. Hence, the model would suffer at the expense of a high computational load.

(ii) Object detection from the video taken by drone via MASK RCNN:

Mask R-CNN [29] is developed on the top of faster R-CNN, which is used for object segmentation, with the similar functionalities in addition to the operations it predicts and identifies class label, object mask, and the bounding box simultaneously in an image.

(iii) Object detection from the video taken by drone via faster RCNN:

Faster R-CNN stands out as a deep convolutional network tailored for object detection, offering users a unified and end-to-end solution. This network excels in swiftly and accurately predicting the positions of various objects within an image. The key innovation introduced by faster R-CNN, as described in reference [30], is the incorporation of RoI pooling layers. Unlike its predecessor, R-CNN, which employed a convolutional neural network (CNN) input for individual regions, faster R-CNN takes the entire image as its input.

(iv) Object detection from the video taken by drone via YOLO:

Yolo stands for "You Only Look Once." This model detects and recognizes multiple objects in an image. This is considered to be the prediction problem, which gives probabilities for the identified object in an image. YOLO [31] is a one-step object detector that extremely improves the computational performance in object detection. Like faster R-CNN, YOLO uses a single feature map to recognize objects and the difference is that the image is divided into a grid and the items are searched within it. The vital part in YOLO algorithm is that it works well for the real-time data in object prediction. It produces accurate prediction with less background noise. These features are possible because YOLO has excellent learning functionality about the depictions of the object in an image and is used for object detection.

(v) Object detection from the video taken by drone via FPN:

Feature pyramid network is considered to be the most important model for accurately identifying objects at various scales. FPN has been advanced multiple times and has undergone significant modification, making it a structural element of contemporary object detectors.

4.9 Deep Learning for Motion Control

Recently, deep learning motion control algorithms have been used in a number of scientific studies. Traditional control model has involved in a large number of robotic complex problems in a quick and accurate manner [32]. This standard control theory, however, only resolves the issue for a particular situation and for an approximate robot model, and is unable to readily adjust to changes in the robot model or when UAV gets damaged, wind drafts, and heavy rain occurs. Based on deep learning techniques, learning from experience will make the model give high accuracy and high performance in their output. Hence, learning from experience is crucial in terms of hardware damage situations and can overcome a number of complex problems that are subject to change over time. There are two types of UAV navigation in motion control as shown in Figure 4.20.

Steps involved in motion control by deep learning are as follows:

- It properly generalizes certain sets of labeled input data.
- It allows gathering a proper pattern or information from raw inputs such as images and sensor data, which can lead to proper behavior even in hostile situations.

Predicted environments:

Predicted environments are said to be the navigable areas that are predicted from difference in the images by bounding box representation. The biggest bounding box in the center represents the proceeding point. These are the fundamental procedures of the UAV flights to perform its operation.

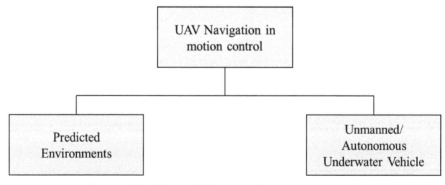

Figure 4.20   Types of UAV navigation in motion control.

Deep learning algorithms handle predicted environments by the following stages:

- A model predictive controller (MPC) is used to produce data at training time.
- In testing time, even in the unknown circumstances UAV is able to track an obstacle free trajectory.
- Inception v3 model, which is a pre-trained CNN can be used in order to enable the output layer.
- After retraining, the UAV accomplished to cross an area occupied with uncommon obstacles in random locations.

Unpredicted environments:

Unmanned aerial navigation in terms of unpredicted environments focuses on cluttered natural environments such as dense forest, tropical forest, hills, and mountain areas [33]. Deep learning algorithms handle unpredicted environments by the following stages:

- Train the model by mapping raw image to number of movement probabilities, which includes turn right, take left, straight path, etc.
- The output is sampled by the activation functions such as softmax, tanh, and relu.
- Testing the prediction by an ODROID-U3 processor.
- Comparing the output with the manual observers.

4.10 Summary

Deep learning models have found a wide range of applications in object tracking, particularly in enabling high-level control when integrated with unmanned aerial vehicles (UAVs). These algorithms have been increasingly used in feature extraction systems, enhancing the capabilities of UAVs. However, the field of object tracking still faces several limitations that require attention. While deep learning has made strides in feature extraction modules, the development of systems at the higher levels of abstraction remains relatively unexplored by the research community. Additionally, systems operating at the lower level of abstraction, like feature extraction modules, face challenges in terms of computational resources. Integrating these resources onboard UAVs is difficult due to the need for robust communication abilities and off-board processing. The current computational resources often fall short, especially when it comes to online processing, which is crucial for applications requiring real-time and reactive behaviors. The limitations in hardware technology emphasize the importance of advancements in embedded hardware to address these challenges effectively. Simultaneously, these limitations should motivate researchers and academics to develop deep learning architectures that are more computationally efficient and optimized for resource-constrained environments.

By addressing these mentioned limitations, there is a significant opportunity to enhance object tracking in high-end applications, especially within the context of unmanned aerial vehicles.

References

[1] D. shah, "Deep Learning | Computer Vision | AI | Open to freelance writing opportunities. |," 26 April 2020. [Online]. Available: https://medium.com/visionwizard/object-tracking-675d7a33e687.

[2] N. Klingler, "viso.ai," [Online]. Available: https://viso.ai/deep-learning/object-tracking/

[3] J Bolme, D.S., Beveridge, J.R., Draper, B.A. and Lui, Y.M., 2010, June. Visual object tracking using adaptive correlation filters. In 2010 IEEE computer society conference on computer vision and pattern recognition (pp. 2544-2550). IEEE.

[4] Shen, J., Liang, Z., Liu, J., Sun, H., Shao, L. and Tao, D., 2018. Multiobject tracking by submodular optimization. IEEE transactions on cybernetics, 49(6), pp.1990-2001.

[5] Kart, U., Lukezic, A., Kristan, M., Kamarainen, J.K. and Matas, J., 2019. Object tracking by reconstruction with view-specific discriminative correlation filters. In Proceedings of the IEEE/CVF Conference on Computer Vision and Pattern Recognition (pp. 1339-1348).

[6] Lu, X., Wang, W., Shen, J., Tai, Y.W., Crandall, D.J. and Hoi, S.C., 2020. Learning video object segmentation from unlabeled videos. In Proceedings of the IEEE/CVF conference on computer vision and pattern recognition (pp. 8960-8970).

[7] Zhang, Y., Wang, T., Liu, K., Zhang, B. and Chen, L., 2021. Recent advances of single-object tracking methods: A brief survey. Neurocomputing, 455, pp.1-11.

[8] Luo, W., Xing, J., Milan, A., Zhang, X., Liu, W. and Kim, T.K., 2021. Multiple object tracking: A literature review. Artificial Intelligence, 293, p.103448.

[9] Zhuo, X., Koch, T., Kurz, F., Fraundorfer, F. and Reinartz, P., 2017. Automatic UAV image geo-registration by matching UAV images to georeferenced image data. Remote Sensing, 9(4), p.376.

[10] Oubbati, O.S., Atiquzzaman, M., Ahanger, T.A. and Ibrahim, A., 2020. Softwarization of UAV networks: A survey of applications and future trends. IEEE Access, 8, pp.98073-98125.

[11] Wu, X., Li, W., Hong, D., Tao, R. and Du, Q., 2021. Deep learning for unmanned aerial vehicle-based object detection and tracking: a survey. IEEE Geoscience and Remote Sensing Magazine, 10(1), pp.91-124..

[12] Arfaoui, A., 2017. Unmanned aerial vehicle: Review of onboard sensors, application fields, open problems and research issues. Int. J. Image Process, 11(1), pp.12-24.

[13] "https://www.e-education.psu.edu/geog892/node/5," [Online]

[14] https://www.gov.uk/guidance/uk-forces-operations-in-afghanistan

[15] https://www.t3.com/features/best-cheap-drone

[16] https://en.wikipedia.org/wiki/Black_Hornet_Nano

[17] https://www.airforce-technology.com/projects/heron-mk-ii-male-uav/

[18] https://en.wikipedia.org/wiki/General_Atomics_MQ-9_Reaper

[19] M. Alzenad, M. Z. Shakir, H. Yanikomeroglu and M.-S. Alouini

[20] Rasti, B., Hong, D., Hang, R., Ghamisi, P., Kang, X., Chanussot, J. and Benediktsson, J.A., 2020. Feature extraction for hyperspectral imagery: The evolution from shallow to deep: Overview and toolbox. IEEE Geoscience and Remote Sensing Magazine, 8(4), pp.60-88.

[21] Ushasukhanya, S. and Jothilakshmi, S., 2021. Real-time human detection for electricity conservation using pruned-SSD and arduino. International Journal of Electrical and Computer Engineering, 11(2), pp.1510-1520.

[22] He, K., Gkioxari, G., Dollár, P. and Girshick, R., 2017. Mask r-cnn. In Proceedings of the IEEE international conference on computer vision (pp. 2961-2969).

[23] Kwak, J., Lee, S., Baek, J. and Chu, B., 2022. Autonomous UAV Target Tracking and Safe Landing on a Leveling Mobile Platform. International Journal of Precision Engineering and Manufacturing, 23(3), pp.305-317.

[24] McEnroe, P., Wang, S. and Liyanage, M., 2022. A survey on the convergence of edge computing and AI for UAVs: Opportunities and challenges. IEEE Internet of Things Journal.

[25] Ecke S, Dempewolf J, Frey J, Schwaller A, Endres E, Klemmt H-J, Tiede D, Seifert T. UAV-Based Forest Health Monitoring: A Systematic Review. Remote Sensing. 2022; 14(13):3205.

[26] Ushasukhanya, S. and Karthikeyan, M., 2022. Automatic Human Detection Using Reinforced Faster-RCNN for Electricity Conservation System. INTELLIGENT AUTOMATION AND SOFT COMPUTING, 32(2), pp.1261-1275.

[27] Ren, S., He, K., Girshick, R. and Sun, J., 2015. Faster r-cnn: Towards real-time object detection with region proposal networks. Advances in neural information processing systems, 28.

[28] J. Redmon, S. Divvala, R. Girshick and A. Farhadi, "You only look once: Unified, real-time object detection.," IEEE Conference on Computer Vision and Pattern Recognition,, vol. 27, p. 779–788, 2016

[29] Carrio, A., Sampedro, C., Rodriguez-Ramos, A. and Campoy, P., 2017. A review of deep learning methods and applications for unmanned aerial vehicles. Journal of Sensors, 2017.

[30] Kamel Boudjit& Naeem Ramzan (2022) Human detection based on deep learning YOLO-v2 for real-time UAV applications, Journal of Experimental & Theoretical Artificial Intelligence, 34:3, 527-544, DOI: 10.1080/0952813X.2021.1907793

[31] Mittal, P., Singh, R. and Sharma, A., 2020. Deep learning-based object detection in low-altitude UAV datasets: A survey. Image and Vision Computing, 104, p.104046.

[32] Golabi, M., Nejad, M.G. (2022). Intelligent and Fuzzy UAV Transportation Applications in Aviation 4.0. In: Kahraman, C., Aydın, S. (eds) Intelligent and Fuzzy Techniques in Aviation 4.0. Studies in Systems, Decision and Control, vol 372. Springer, Cham.

[33] Kim, Sung-Geon, Euibum Lee, Ic-Pyo Hong, and Jong-GwanYook. 2022. "Review of Intentional Electromagnetic Interference on UAV Sensor Modules and Experimental Study" *Sensors* 22, no. 6: 2384.

5

Deep Learning Models for Urban Aerial Mobility: A Review

K. Kathiresan[1], N. Yuvaraj[2], Sri Preethaa K. R.[2], and Sangwoo Jeon[3]

[1]Department of Computer Science and Engineering (AIML), Sri Eshwar College of Engineering, India
[2]School of Computer Science and Engineering, Vellore Institute of Technology, Vellore, India
[3]Department of Computer Science and Engineering, Konkuk University, South Korea
E-mail: kathiresan.k@sece.ac.in; yuvaraj.n@vit.ac.in; sripreethaa.kr@vit.ac.in; jswp5580@konkuk.ac.kr

Abstract

DNNs (deep neural networks) have built significant advancements in handling images, time series, audio, video, and numerous other categories of data by learning representation from data with a tremendous capacity. To assemble the volume of knowledge generated in the remote sensing field's subfields, analyses and literature reviews clearly concerning DNN algorithms' applications have been carried out. Applications based on unmanned aerial vehicles (UAVs) have recently subjugated aerial sensing investigation. Our work aims to give a thorough analysis of how deep learning (DL) principles were used in UAV-based photography. The primary goal is to describe the classification and regression methods applied recently to data collected by UAVs. During this review investigation, the international scientific journals' databases were inspected. This study assessed the published materials' application, sensor, and technique qualities. We discuss how DL shows promise and can handle jobs involving the processing of UAV-based image data. We conclude by speculating on possible DL directions to be investigated in the UAV remote sensing sector. This review aims to introduce, comment on, and review the most recent UAV-based picture applications using DNN algorithms in various remote sensing subfields, classifying them according to their relevance to the environment, cities, and agriculture.

Keywords: Urban aerial mobility, deep learning, computer vision, autonomous system

5.1 Introduction

Drones fly without a pilot on board; unmanned aerial vehicles (UAVs) are these types of aircraft. Given the considerations of aircraft prices, pilot safety, training requirements, and sizes, UAVs offer numerous benefits that greatly expand their range of potential applications. The statistics for the UAV business are excellent. Value market research estimates that the vertical takeoff and landing (VTOL) UAV market will be worth over $10,163 million by 2024 [1]. In reality, several civil and military application fields have already witnessed unprecedented UAV growth [2]. UAVs can fly independently or under remote control from a pilot. In the first scenario, the pilot is at a ground control station (GCS) for human control that is either on land or at sea. Deep learning is suitable for a wide range of autonomous robotic applications, as seen by the recent rise in deep learning robot-related research articles, which is expected to continue growing [3].

UAVs' adaptability, automation potential, and low cost have led to a sharp surge in civilian applications in various industries during the past several years. Inspecting electricity lines [4], conserving animals [5], inspecting buildings [6], and practicing precision agriculture [7] are a few examples. Situational and self-awareness are obtained by exteroceptive and proprioceptive sensors, respectively [8]. Although both have significant shortcomings, UAVs have been fitted with sensors that give location and orientation in space, such as the inertial navigation system (INS) and the global positioning system (GPS). Additionally, urban canyons and indoor routing can substantially impair GPS accuracy. The accuracy of the GPS is dependent on the total amount of satellites that are currently in orbit. Comparing UAV-enabled communication systems to conventional communication systems, the former may offer a higher level of service and is more economical. UAVs have been used in various communication systems to improve line-of-sight communication links because they can be cost-effectively and flexibly deployed and move about their surroundings quickly. Academia and business have paid close attention to using UAVs as new data-collecting platforms [9]. Platforms that support UAVs provide several benefits over traditional data-collection methods.

Urban aerial mobility (UAM) uses the air to increase connectivity between people and places by strengthening links between them and the cities and regions in which they live. A multimodal mobility system can benefit from UAM's contribution. Sustainable city growth is made feasible when urban transportation takes to the skies and loved ones are no longer as far away as they once were. Around 60% of people on Earth will live in urban areas by 2030. Since ground infrastructure will become increasingly crowded due to this enormous population expansion, there will be a tremendous demand for cutting-edge transportation solutions. One approach may be to give people access to a reliable, practical, and secure method that uses airspace over densely populated areas.

5.2 Background Study

5.2.1 UAV applications

The language used to discuss unmanned aerial vehicles (UAV) is diverse and occasionally challenging to understand. The first UAVs were used in the military, where two technical terms were used to describe them: (1) A UAV is an aircraft carrying all its cargo on board; (2) unmanned aircraft system (UAS), which assembles the entire relevant parts, comprising the UAV itself, the ground section (i.e., the ground control station), and the interaction section. The military continues to use these expressions [10], and the civilian world is likewise well-known and uses them frequently. To distinguish them from military UAVs or UAS and to emphasize the role of a remote pilot responsible for the legal duty of flying UAVs, the terms remotely piloted aircraft (RPA) and remotely piloted aerial system (RPAS) were first introduced in civilian applications. These phrases are, nonetheless, no longer relevant. Since it highlights that it performs an autonomous action, researchers in robotics frequently use the name "aerial robot" [11]. The human remote pilot's function is just that of a supervisor, taking over control in the event of an unexpected and irrecoverable failure.

There are various distinct types of UAVs, and they may be categorized using several factors [12], including, among other things, the lifting mechanism, range, maximum takeoff weight (MTOW), endurance, shipment weight, size, height, operational circumstances, or degree of autonomy. The lifting technique, which differentiates lighter than air, flapping wing, fixed-wing, hybrid designs, and rotary-wing, is a very extensive criterion for categorization. The most common UAV configurations are fixed-wing and rotary-wing. While rotary-wing UAVs benefit from hovering, VTOL, and short-speed flights at the cost of a smaller endurance and range, they are ultimate for an even greater variety of applications. Fixed-wing UAVs are typically utilized for high-level altitude, extreme range, and long-endurance applications.

5.2.2 Sensors onboard UAVs

UAVs today are capable of holding a wide variety of sensors and cameras. UAVs are versatile and adaptable for a wide range of high-performance applications because of their downsizing and cost-effectiveness on the one hand and their smart sensors and high-resolution cameras on the other. The technology employed will vary depending on the dimensions of the UAVs and the kind level of data that has to be gathered. The variety of modern sensor and imaging technologies that may generally be put on UAVs comprises GPS, INS, standard cameras, multispectral cameras, hyperspectral thermal sensors, LiDAR and radar sensors, and numerous more specified sensors. UAVs have sensors on board to gather data about the surroundings (exteroceptive) or the UAV itself (proprioceptive). Their usage is influenced by various variables, including the application, the environment, size, and UAV cost; the tasks that need to be completed; the degree of safety; the degree of the UAV, etc. Every sensor has a unique set of functional properties, which gives each sensor a unique set of benefits and drawbacks.

An RGB camera is a passive sensor that captures visual information in the visible spectrum of light within the observed area using three separate channels: red, green, and blue (RGB). When mounted on a UAV, the primary challenges include the vehicle's speed or sudden movements, which can result in image blur and noise. To address this, global shutter cameras are often employed. Unlike their rolling shutter counterparts that scan the scene either vertically or horizontally, global shutter cameras capture the entire image simultaneously [13]. On the other hand, a gray-scale camera captures the average intensity of the visual spectrum of light in the scene using only one channel.

Thermal cameras are passive sensors that catch infrared radiation from things warmer than absolute zero (thermal radiation). Many applications were made possible when their price recently dropped [14]. However, their initial employment was only for military night vision and surveillance. In reality, thermal imaging may solve the lighting issues with conventional gray-scale and RGB cameras, offering a precise answer in all fields requiring the detection and tracking of life, and UAVs are no exception [15].

Event-based cameras, such as the dynamic vision sensor (DVS), are particularly suited for real-time motion analysis because they have low latency, significant sensitivity to light, and increased temporal resolution, deployed in UAVs [16]. Since the output is not built using traditional intensity imaging, but rather a series of asynchronous events, more study and developing alternative image processing pipelines are urgently needed [17].

Active sensors called LiDARs (light detection and ranging) measure distances by projecting laser light to a target and reflecting with a transducer. In the past, LiDARs were big, expensive, and heavy to be used on UAVs. However, recent breakthroughs in solid-state technology, with standards like Ouster LiDARs, have significantly lowered their size, weight, and price, making them a more and more attractive option for users of UAVs.

5.2.3 Autonomous operation of UAVs

UAVs must operate to increase their effectiveness, dependability, scalability, security, and safety, facilitate new applications, lower operating costs, and conduct autonomously 4D-based tasks (dull, dirty, dangerous, and dear). However, obtaining the utterly autonomous operation of UAVs is still a research issue that has to be resolved. Several studies are concentrating on adaptable designs for aerial robotic systems to accomplish utterly autonomous operations, such as the widely utilized Aerostack [18] (see Figure 5.1). These designs, however, rely on the availability of many pre-assembled parts with well-defined functions, which typically need more functionality, operation, and maturity.

Situation awareness is the most crucial and difficult among these components. Using the data collected by the sensors (mainly those on board), this component oversees perceiving the elements in the surroundings to provide a depiction of the surroundings and the UAV's condition over time and space [19].

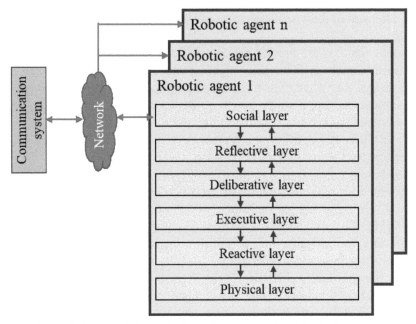

Figure 5.1   Aerostack architecture of an unmanned aerial robotic system.

In many circumstances, the raw data produced by the sensors need to be simplified to be used directly in situational awareness procedures. They must instead be distilled down by removing higher-level information. The feature extraction components perform this pre-processing step [20].

We have organized the papers using the taxonomy suggested in Aerostack [21], an aerial-robotics-architecture congruent with the typical elements associated with guidance, observation, routing, and control of unmanned rotorcraft systems. To better understand the components of the investigated aerial robotic systems, it is necessary to refer to this design, which is shown in Figure 5.2. The elements that make up an unmanned aerial automated system, according to Aerostack, may fall into procedures and interfaces:

(i) Hardware interfaces: It consists of sensors and actuators.

(ii) Motor system: Motion controllers are components of a motor system, and they typically receive commands containing desired values for a variable such as position, orientation, or speed. These desired values are then converted into low-level instructions that enable communication with actuators.

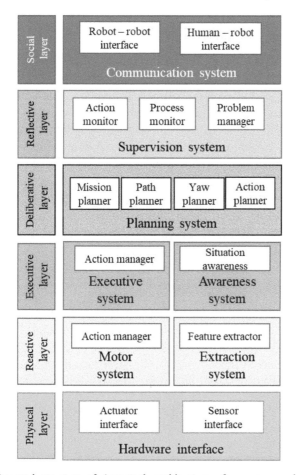

Figure 5.2 Layered structure of Aerostack architecture of an unmanned aerial robotic system.

(iii) Feature extraction system: In this context, feature extraction refers to the process of extracting relevant characteristics or interpretations from sensor data. Deep learning algorithms, in particular, have the inherent ability to learn data interpretations as their primary function, which makes feature extraction techniques inherently present in deep learning algorithms.

(iv) Situational-awareness system: It has elements that combine sensor data into variables that describe the robot and its surroundings to comprehend the environment. SLAM algorithms are an illustration of a situational awareness system component.

(v) Executive system: It builds intricate behavior sequences from receiving high-level symbolic actions.

(vi) Planning system: This approach uses planning to create comprehensive answers to complex problems (mission and path planning).

(vii) Supervision system: A component of the supervisory system can oversee other integrated systems, simulating self-awareness in this manner. This sort of component may be shown using an algorithm that determines if the robot is genuinely moving toward its objective and responds to issues (unexpected barriers, errors, etc.) with retrieval procedures.

(viii) Communication system: Establishing effective communication with human operators and other robots is the responsibility of the communication system's components.

5.2.4 The cognitive cycle of UAM

In modern systems, autonomy (from ancient Greek: auto-"self" + nomos-"law") is a desired quality. Autonomy is the capability to adjust to variations in the environment or oneself. Autonomous systems can be built on various architectures with increasing levels of sophistication, from straightforward reactive architecture to complex deliberative design to cognitive architecture. Simple Sense-Act loops are the foundation of reactive systems, but more complex Sense-Decide-Act loops are used in deliberative systems to provide the system with the ability to think and make decisions. The Sense-Aware-Decide-Act-Adapt-Learn cognitive cycle [22] is a five-step process that may be used to create more complex autonomous systems. This cycle outspreads the deliberative process of Sense-Decide-Act by including condition awareness, adaptability, and learning competencies, as shown in Figure 5.3. These systems collect data from several hard and soft data sources in amounts comparable to the Internet's total volume. Soft data comprises human input, virtual-sensing, fused, and social-media data. Complex data is gathered from actual physical sensors (such as cameras, radar, and LiDAR). The collected information will be processed to reach several degrees of state consciousness: the perception of environmental variables within particular time and space restrictions, the interpretation of their significance, and the projection of their state shortly and probable implications [23]. The result of perception is a collected image of the world, which may contain various sub-processes such as object recognition, object localization (or tracking), and object identification.

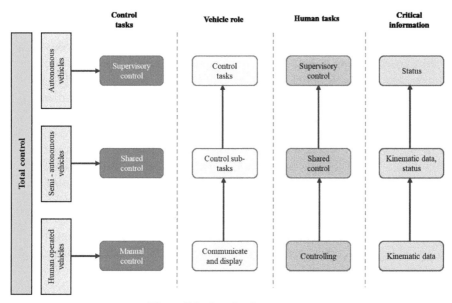

Figure 5.3 Levels of autonomy.

The capacity to comprehend combines this data to create a complete picture of the world that demonstrates the connections between the identified things. Future activities of objects in the environment are projected via projection capability. The turnout of the condition awareness capability is typically a domain model or mapped environment that may include an initial state, goal state, potential robot states, fundamental movements or operations, rigid and flexible limitations, ambiguity (as related to sensors and actuators), previous information, and a static or dynamic map with low or high resolution.

In the need for clarity, decisions are made through decision-making modules. If this uncertainty is managed correctly, it may lead to correct assumptions about its condition and the state of the environment, which might lead to incorrect actions being taken. The system can modify its behavior using an adaptation module depending on a context taken from the situation-awareness components. The detailed data responds to situation-related inquiries like what, when, where, etc., regarding activity, location, status, and time. The modification can take the form of altering job descriptions, the coordination method, or re-designing and controlling the use of resources in a way that progresses the data collection process. For autonomous systems, the ability to learn from past experiences is essential. A learning module may be created to enable the system to pick up new skills or automate tedious jobs while still offering intelligent instruction to a person if necessary. Motion planning is a challenging task that chooses an autonomous robot's future path of actions or activities to move from its starting point to a predetermined objective state.

Interestingly, the Robotics, Autonomous Systems Roadmap, and Tele-Robotics working group at NASA distinguished between automated and autonomous systems [24]. According to their definition, an automated system is one that "follows a script, although one that is potentially highly intelligent; if it reaches an unexpected scenario, it breaks process and stays for human assistance". The decisions have already been made and are encoded in some fashion or will be made independently of the method. An autonomous system makes decisions on its own. The system is independent of the individual on whose behalf the objectives are being accomplished since the purposes it attempts to reach are supplied by another entity. Although the decision-making procedures may be straightforward, local decisions are still made.

According to the degree of autonomy, cars and robots may be divided into three categories: semi-autonomous, human-operated, and completely autonomous. In the vehicle control process, system roles, human activities, and critical information are observed at each level [22]. Figure 5.3 highlights these three levels and their relationship to the control taxonomy.

The control level in Figure 5.3 establishes the relative priority and the jobs humans perform. This indicates that human accountability for a vehicle's operation ranges from complete control to strategic control. The human is in absolute control during total control and is in charge of all choices, including trajectory control and strategic planning. On the opposite end of the spectrum, the human is solely in order of reasonably long-term goals, at least while the vehicle is doing the work.

- **Manual control**: At this level, all robot operations are entirely under the direction of the human operator; the robot's job is to show operational and contextual data and respond to user interaction.
- **Manual control with intelligent assistance**: With the development of increasingly intelligent robots, a person can educate a robot on the basics of a job site, including specifying areas that should not be entered. The robot may change information displays and alter human inputs to give instructions. The ability of the human to delegate more responsibility to the robot increases as its capabilities increase.

- **Shared control**: Certain subtasks are under human control at this level, while others are under simultaneous robot control.
- **Traded control**: Subtasks are alternately handled by a robot and a person.
- **Supervisory control**: At this stage, the human monitors the robot's performance of duties and occasionally steps in to help in an emergency. The human user comes into the control loop only when abnormal circumstances occur. Other autonomous vehicles based on the cognitive cycle include river taxis, air taxis, autonomous ships, and surveillance or search and rescue drones.

5.3 Deep Learning in the Framework of Machine Learning

Machine learning (ML) is a capability that enables artificial intelligence (AI) systems to learn from data. The following sentence provides a concise explanation of the learning process: "Regarding a class of tasks T and a designated performance metric P, a programming system is said to learn from experiences E if its performance on tasks within T, as measured by P, improves with accumulated experiences E" [25]. The classification of ML algorithms into the three categories of unsupervised, supervised, and reinforcement learning often depends on the nature of these accumulated experiences E.

Given a training collection of input and output samples, supervised learning algorithms study how to link input with an inevitable outcome [26]. Nowadays, the most critical techniques in supervised learning are FNNs, CNNs, RNNs, and LSTM models for RNNs. Feed-forward neural networks are the most popular supervised learning models, often known as multilayer perceptron (MLP).

CNNs are a particular class of models designed to handle two-dimensional input data, such as pictures or time series data, as shown in Figure 5.4. These models are named after the linear convolution mathematical technique, which is always present in at least one network layer.

In deep learning, 2D convolution image I with a kernel K is the highly used convolution operation and is signified by the following equation:

$$C(x, y) = (I * K)(x, y) \tag{5.1}$$

$$= \sum m \sum n \, I(m, n) K(x - m, y - n). \tag{5.2}$$

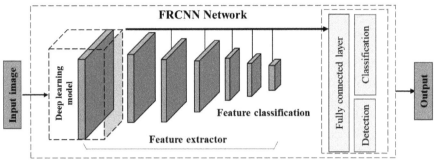

Figure 5.4   A Convolutional neural network model.

The result of the convolution process is often subjected to a nonlinear activation function before being further altered by a pooling function, which substitutes a value for a particular output position with values derived from surrounding outputs. This pooling function conducts subsampling of the input data and aids in making the representation learned invariant to tiny translations of the input.

Recurrent neural networks are models that, in contrast to MLPs, have an output that depends on the current inputs and the prior outputs, which are stored in a hidden state h. As a result, RNNs may encode the data in the sequence by itself, but MLPs need a recollection of the prior outputs. Learning from sequential data may be accomplished using this kind of model. RNNs are often trained to utilize backpropagation through time (BPTT), a backpropagation extension that incorporates temporality into the computation of the gradients. Many problems arise when using this strategy with lengthy temporal sequences. Gradients accumulated over a long sequence are either extraordinarily big or immensely small.

Long short-term memory (LSTM) models are a form of recurrent neural network (RNN) architecture established in the year 1997 by Hochreiter and Schmidhuber [27], which successfully solves the problem of vanishing gradients by using gated cells to maintain a more consistent error, allowing for continuous learning across a more significant amount of steps. A primary LSTM cell is illustrated in Figure 5.5.

Along with RNNs' outer recurrence, LSTM-gated cells in RNNs also exhibit interior recurrence. You can write to and read from the internal state that cells store.

Unsupervised learning aims to create models that draw high-level, meaningful representations from highly dimensional, unlabeled sensory data. The

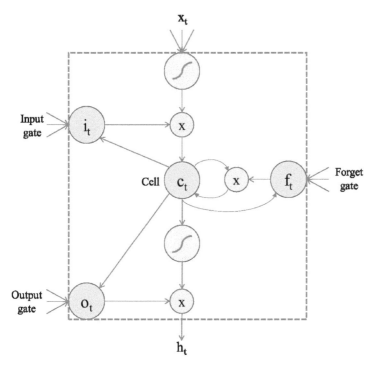

Figure 5.5 A long short-term memory model.

visual cortex, which requires a relatively tiny quantity of labeled input, inspires this functionality. Unlike supervised learning, unsupervised learning algorithms enable users to perform more complicated processing tasks. Unsupervised learning, however, can be more unpredictable than other types of spontaneous learning. Algorithms for unsupervised learning include neural networks, anomaly detection, clustering, etc.

An unsupervised neural network, known as an autoencoder (Figure 5.6), utilizes pre-set target values to be identical to the input data. The autoencoder consists of two primary networks: the "encoder," which transforms the input data into a low-dimensional code, and the "decoder," which reconstructs the original data from the code. During training, the objective is to minimize the error between the original data and its reconstructed version by fine-tuning these deep models.

In reinforcement learning, an agent interacts with the natural environment to choose the optimum course of action for each state at any moment (see

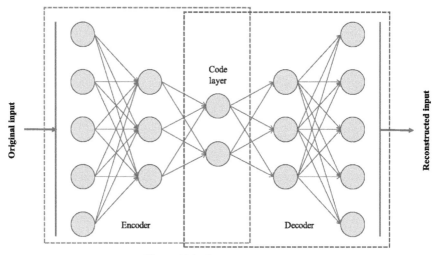

Figure 5.6   Deep autoencoder.

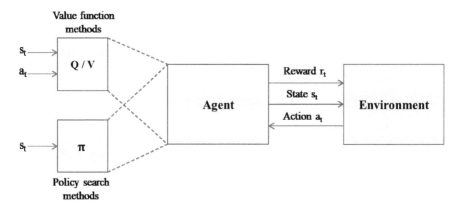

Figure 5.7   Generic structure of a reinforcement learning problem.

Figure 5.7). The agent must balance exploring and exploiting the state space to decide the best course of action that increases the accumulative benefit from connections with the environment. In this situation, an agent adjusts its behavior or policy while conscious of the current conditions, actions, and rewards at each interval. An optimization method is created throughout the entire state space to optimize the total reward.

5.3.1 Deep learning models for UAM

Urban air mobility (UAM) seeks to revolutionize travel in the airspace, similar to how autonomous vehicles (AVs) aim to do so on the road. Researchers and industry leaders envision a future where conventional transportation extends to the low-altitude sky rather than solely focusing on improving driving efficiency or increasing road capacity [60, 47]. As previously mentioned, simulation and modeling have firmly established their role in autonomous vehicle research, while UAM has received comparatively limited attention from the simulation and modeling community.

The industry standard for regional and municipal planners in creating a comprehensive travel demand, land use, and transportation model [87, 84] for urban air mobility (UAM) adopts a holistic approach. While basic templates from autonomous vehicles (AV) exist to understand the implications of emerging modes on cities, there is a need for further expansion [71, 86]. Specifically, simulations should enable assessments based on UAM network designs, acceptance rates, and various urban air taxi configurations, as well as optimal operations considering specific constraints, such as reducing overall trip times and emissions [87]. Disaggregate action-based models are increasing in popularity and might be helpful in achieving this aim, even though aggregate techniques have historically been utilized in travel demand modeling [85]. A person's or a household's daily activity pattern may be developed and used as input into simulation models since decisions like mode selection depend on residential location and employment location choices [69, 85]. Mesoscopic models and MATSim were used in the first research to examine the UAM systems in Sioux Falls and Munich, although effect assessments should be noticed more [71, 72].

5.3.2 Understanding the potential impacts of UAM

This section examines possible effects, factors to take into account, and outstanding issues in relation to the following subject fields: (a) transportation and usage of land, (b) equity, (c) safety, (d) noisy environment, and (e) sustainability.

(a) Transportation and usage of land:

The potential effects of urban air mobility (UAM) passenger services on travel patterns and land usage have been estimated through various assessments. Many of these investigations are based on four key hypotheses. Firstly, considering the high cost of airplanes, most of these analyses assume that UAM would operate as a shared platform [81]. Secondly, similar to transportation network companies (TNCs), these studies generally assume that customers would require cell phones for access [55]. Thirdly, according to the predictions in these studies, VTOL aircraft would primarily land and take off at vertiports, which are landing facilities designed for urban areas, resembling runways and landing pads [47, 60]. Fourthly, several analyses suggest that initial UAM operations would adopt an "air-metro" operating model involving flights along specific high-volume transit corridors connecting a sparse network of vertiports [79].

These premises enable a more thorough analysis of the impact of UAM on travel patterns and urban design. Emerging mobility technologies (AV and UAM) aim to address the identified problems with different choice and third-party logistics processes, namely trip period, a theoretically recognized "poor" component that promotes disutility [76, 50]. This issue is addressed in various ways by both AV and UAM. UAM aims to reduce actual travel time, whereas AV aims to reduce the perceived trip period by replacing driving time with more activities like working or resting. It reduces the overall expense of travel, and UAM's strategy may be straightforward. The predicted reduction in overall travel expenses brought on by UAM may alter travel habits in several ways.

First, traveling longer distances or transferring commodities over 30 miles may be switched to aerial methods, and the overall VMT may drop on the ground or in the air [81]. The circuity factor, which represents the ratio of distance traveled on land to the air for an origin−destination (OD) pair, typically exceeds 1.5 in the United States and reaches around 2 in other countries worldwide [57]. This suggests a potential decrease in the total distance traveled. However, this outcome depends on three key considerations. One concern is the strategic placement of vertiports to minimize the first-mile/last-mile connections to the vertiports themselves while accommodating a higher number of passengers, logistics, and emergency services that could utilize urban air mobility (UAM).

According to the bid-rent model discussed earlier, if the average travel cost decreases in any way, there is likely to be a shift away from the central business district (CBD), leading to urban consolidation. However, if the United States expects 500 million UAS deliveries and 750 million passenger journeys by 2030, automotive infrastructure, especially parking space and road capacity, may decrease [79]. Instead, urban mixed-use infill construction could prioritize factors such as transit-based development, and the integration of public transportation stops with private and public transportation pathways. Making these highly desirable regions with high land values more accessible may encourage further residential clustering [54]. Infill development is another option for lower-density neighborhoods to provide more accessible places close to their nearby vertiports [43, 45, 35], which may lead to both uniqueness and scattering. Accordingly, the urban structure will be more polycentric and spread on a macro range. At the same time, each center could have higher densities and accessibilities close to its regional transport stations and vertiports [28]. Higher VMT and higher car emissions are two problems associated with macroscopic sprawl [34]. If the planes are not as electric as an AV [30, 44], the sprawl that may ensue could also harm the environment.

(b) Equity:

Several equity issues have been brought up concerning UAM, the most being affordability and scalability. Many people are worried that UAM will only be available to wealthy homes as a passenger use case [41, 53]. In places like New York, Mexico City, San Francisco, and Sao Paulo, on-demand passenger services provided by helicopters are high-end businesses that cater to a mostly wealthy clientele [83]. While scaled operations, aircraft autonomy, and electrification have the potential to broaden the market demographic for urban air mobility (UAM), the extent to which the mass market would be willing and able to pay for these services beyond early market studies and various assumptions is still being determined [79]. Even in the most favorable circumstances, achieving widespread market adoption of urban air mobility (UAM) may not be feasible as it could still be too expensive and challenging for older individuals, low-income families, and underserved populations to access.

UAM could also have two additional equity-related effects. Firstly, as discussed in the previous section, UAM has the potential to both increase and decrease travel demand depending on its impact on the surface transportation network. In such a scenario, low-income travelers who heavily rely on their vehicles or public transit may bear a disproportionate burden of the costs associated with the induced demand, resulting in increased traffic congestion and longer travel times [52, 61]. Secondly, UAM may directly compete with commuter rail or long-distance public transit, leading to reduced demand, increased costs, and decreased service for these existing transportation modes. Minority communities and individuals from disadvantaged backgrounds, who constitute a significant portion of transport users in many cities, may also be disproportionately affected by these changes [58].

Additionally, there are several ways in which the installation of vertiports might harm vulnerable communities. Vertiports may be erected in low-income or underserved areas, which would have an adverse effect on the environment in terms of undesired traffic, noise, and other environmental effects [52, 39, 81]. As a result of lesser demand, vertiports may alternatively be placed outside of low-income centers, discouraging low-income families from using them and resulting in an unfair feedback loop.

The impact of urban air mobility (UAM) on equity and the strategies to address it varies significantly, and the deployment of equity initiatives continues to face significant challenges. The location of infrastructure, scalability, and affordability can all have adverse effects on underprivileged areas. Suppose UAM remains limited to niche services for either product delivery or passenger transport. In that case, certain areas may never reap its benefits and, even worse, may suffer from its negative externalities, such as increased noise and congestion if vertiports are situated nearby and emissions if early VTOL aircraft are not electric [63, 66, 40]. In the scenario where UAM fails to achieve substantial growth or vertiports are positioned far from disadvantaged neighborhoods, rendering it essentially irrelevant to them, the allocation of public resources toward UAM infrastructure development may have been diverted from mobility initiatives that could have otherwise benefited these communities, such as enhancing bus frequency and reliability [65].

Conversely, if vertiport locations are accessible, UAM prices become affordable, and efforts are made to reduce or eliminate aircraft noise and pollution, UAM may have positive, equitable effects. Approximately 250 businesses in the UAM market prioritize aircraft development to minimize noise and emissions while lowering costs. However, the selection of vertiport locations under equity considerations remains an area of study that has yet to be thoroughly explored.

(c) Safety:

The implementation of urban air mobility (UAM) encounters a significant barrier in terms of safety. Planners and policymakers must establish legal and regulatory frameworks to mitigate safety hazards, as airspace users and the general public express concerns about aircraft safety [81, 47, 28]. Numerous studies have found that the general public often associates UAM with general aviation, light aircraft, and helicopters, which have relatively poorer safety records [62]. From 2003 to 2013, general aviation (GA) reported

7.5 disastrous incidents per 100 million miles flown, compared to 1.3 and 0.068 for commercial aviation and vehicles, respectively [63]. To address this perception of safety risk, UAM aircraft must not only meet or exceed the safety levels achieved in commercial aviation but also overcome the negative association with general aviation and helicopters, especially considering their operation above densely populated areas where the risk is higher [66, 81, 67, 63]. The public often focuses on the number of fatalities per accident in aviation rather than the overall number of annual fatalities or the comparative safety of aircraft compared to other modes [42, 64]. Therefore, UAM criteria may need to dispel this impression. Successful small-scale demonstrations are necessary to expand society's and users' awareness of safety before mass market adoption [74].

Human mistakes, bad weather, and an absence of precise sense-and-avoid skill in the air are just a few of the significant genuine obstacles standing in the way of an impeccable safety record [44, 70]. 80% of general aviation accidents result from pilot mistakes [63]. In order to reduce safety hazards, Nneji et al. [66] add that pilots in manned UAM must preserve safe departure, vehicle management, vehicle control, and mission. Unexpected wind gusts can cause the feedback control systems in both manned and unmanned urban air mobility (UAM) to undercorrect or overcorrect upon detecting the force, resulting in unpredictable behavior [68]. Techniques such as unconventional deep learning, sensing, and real-time visual data collection are still in their early stages and play a crucial role in developing sense-and-avoid technology for unmanned UAM [66]. Concerns also arise regarding unmanned, autonomous aircraft, as people worry about the possibility of them crashing to the ground.

There are two distinct viewpoints on safety: that of the over own and that of the passenger in UAM aircraft. On board the aircraft, passengers must be guaranteed their safety, most likely at a level comparable to that of commercial flying. The over owner must be confident in the stability of the aircraft and its minimal likelihood of colliding with anything while flying over their residences [51]. At many decision-making levels, legal and regulatory frameworks are presently being built. Examples include NASA's control of unmanned air traffic and the FAA and EASA's certification of drones and pilots in Europe [31, 63, 67].

(d) Noisy environment:

The noise generated by urban air mobility (UAM) will pose a significant barrier to its implementation. Multirotor vertical takeoff and landing (VTOL)

UAM aircraft are particularly prone to noise due to the interaction between each rotating blade and the air previously disturbed by the preceding blade [30, 36, 48]. UAM presents two potential scenarios: (1) occasional flights of a single aircraft with noticeable noise that quickly dissipates, or (2) a continuous cacophony from numerous aircraft flying for several hours throughout the day. Addressing this issue is crucial in aircraft and rotor projects, as well as in determining capacity placement and airspace constraints, to gradually reduce the sound exposure level (SEL) under various circumstances [78]. Professionals have set SEL targets below 60 dB at heights up to 300 feet for cargo transportation in smaller aircraft [73].

However, besides volume, there are other factors to consider. The type of noise, in addition to loudness, is a concern for communities affected by excessive noise, as indicated by research conducted in Los Angeles, Mexico City, Switzerland, and New Zealand [88]. Respondents in these studies considered the buzzing sound of bees as a standard for acceptable noise, as it is less intrusive both in terms of loudness and perceived friendliness [88]. However, preferences may vary among different communities [49, 88].

Flight altitudes continue to be a critical factor affecting SEL. As altitude increases, SEL decreases; however, operating at higher altitudes above 10,000 feet can significantly deplete battery power during takeoff and landing operations, necessitating UAM aircraft to operate within confined areas. Flying at higher altitudes may also interfere with class A commercial aircraft, which is unacceptable [77]. Class B, C, and D airspaces are thus expected to be utilized in the vicinity of busy airports and over residential areas. However, because they are located at lower altitudes (between 300 and 3000 feet), they are substantially louder for people on the ground [77]. As a result of restrictions on helicopter routes due to noise issues, many cities have fragmented municipal, state, and federal legislation and guidelines. Communities may need to adapt their aircraft designs and operations to this regulatory environment to integrate UAM in low-level airspace [79, 77].

Because of this, the various legal and regulatory systems within diverse communities will also provide difficulties from a technical and planning standpoint. The idea of "restricted flight plans" refers to the ability of communities to set noise restrictions when airspace users create flight plans [60]. Operations may become more complicated, posing problems with battery life and time savings if UAM flight tracks must dodge noise-sensitive land uses (instead of choosing the shortest path). Constrained flight plans are complicated since various populations, even those close to one another, have different noise sensitivities [82].

(e) Sustainability:

As multimodal mobility spreads, it is necessary to unbiasedly assess the sustainability consequences of each mode to foresee future deployment. Transportation generates the most significant percentage of any industry in the United States, 28% of all greenhouse gas (GHG) emissions. 83% of GHG emissions in the transportation sector come from cars and trucks [80]. While EV and micro-mobility sustainability research are ongoing, UAM's status is still being determined, and its research program is more complicated. Regarding UAM's sustainability, three key inquiries must be made: (1) Do UAM planes produce more significant pollutants? (2) If UAM increases sprawl, would this increase emissions? (3) Will UAM have any indirect effects or procedures that increase emissions, such as deadheading? These three queries will be answered in this section.

Aircraft emissions: Electric vertical takeoff and landing (eVTOL) certification is something that several manufacturers hope to achieve in the 2025s. If an environmentally friendly electric grid powered these planes, they might be emission-free, but lifecycle emissions are frequently overlooked in calculations like this one. Several aircraft industrialists have also started investigating hybrid-power-driven aircraft designs, especially to facilitate power-intensive tasks like hovering [62, 57]. For specific use cases that may be one-off or inherently have smaller ecological footprints, researchers are also exploring the possibility of using internal combustion engine (ICE) vertical takeoff and landing (VTOL) aircraft [28].

Different distribution scenarios for aero planes exist in the future.

(1) eVTOL
(2) eVTOL, hybrid VTOL
(3) eVTOL, hybrid VTOL, ICE VTOL

The amount of UAM aircraft releases is influenced by the likelihood of these eventualities and the proportion of each mode in each scenario. Expected use cases have a significant impact on scenario likelihoods and mode shares. Due to its ability to include a range of use cases in customers and goods UAM, Case (3) may be the most plausible. For instance, according to [28], it was discovered that internal combustion engine (ICE) vertical takeoff and landing (VTOL) aircraft are better suited for activities that are less common but require longer flying periods and greater versatility, such as rural emergency response [28]. In the use case of customer urban air mobility (UAM), electric VTOLs (eVTOLs) emit fewer greenhouse gases (GHGs) per vehicle kilometer traveled (VKT) than internal combustion engine vehicles (ICEVs) for trips exceeding 35 km. However, it should be noted that the average commute distance by ground transportation is only 17 km, and journeys exceeding 35 km account for only 15% of all vehicle journeys. This suggests that there may be limited demand for customer facilities utilizing eVTOL or hybrid VTOL aircraft [57].

Sprawl: According to location theory in urban economics, people tend to migrate farther out from the center as transportation costs decrease [52]. As a result, if customer UAM is widely used for long-distance travel, it is expected that the urban border will grow as residents and employers relocate, resulting in polycentric urban sprawl [29, 75, 38]. If both freight and passenger vehicles do not transition to cleaner fuel sources, this urban sprawl could lead to even higher transportation costs and potential emissions throughout suburban hubs [80]. However, despite the additional sprawl, there may be sustainability benefits if light freight and cargo delivery adopt urban air mobility (UAM) solutions within smaller regions to replace inefficient last-mile delivery vehicles [37]. Alternatively, at the local level, increased urban development and the promotion of activities that support active modes of transportation can enhance accessibility near potential vertiports, which would also contribute to sustainability [33].

Indirect effects: Several other repercussions might happen depending on the circumstances and use case for UAM. Suppose passenger UAM is widely implemented and successfully substitutes long-distance drivers. In that case, the new perceived improvement in the level of service on the road may stimulate additional requests for the automobile, which would then result in an overall rise in congestion and emissions [46, 59]. UAM must reach a certain depth to actualize an induced demand effect, although this depth is still unknown.

Future and current pandemics like COVID-19 may have an impact on passenger UAM. The COVID-19 mandate that many workers must work remotely has resulted in a global decline in air pollution and emissions. It is still being determined how much work will be done remotely in the future. If this ratio is large, congestion may completely disappear due to a drop in travel requests. However, the percentage of remote work stays low. In that case, traffic may rise if the metro region significantly relies on public transportation, which may drive people to UAM [56]. Depending on the kind of aircraft, this modification will have a different impact.

5.4 Discussion

Although each category has demonstrated its unique characteristics, it is necessary to make certain considerations that are common to the many taxonomies examined in this text.

The development of deep learning has shown remarkable results for real-world applications, even with modest end-to-end direct pipelines, as opposed to traditional computer vision and machine learning approaches that often required complex handcrafted engineered solutions to achieve satisfactory results. Two main factors have likely contributed to these advancements. Firstly, the release of object detection datasets relevant to urban air mobility (UAM) research has provided common benchmark platforms and enabled the training of deep networks with large collections of labeled data. Secondly, general-purpose 2D object detection network architectures have evolved, incorporating ideas from classical computer vision techniques.

As a qualitative trend, approaches still rely on traditional object descriptors in UAMs with extremely limited processing resources and vision-based control scenarios. Deep learning is preferred in more expensive solutions or when image processing is performed off-board and/or offline. However, even when used to enhance UAM obstacle avoidance capabilities, these models can have high computational costs. It has been demonstrated that accurate outcomes for UAM obstacle avoidance can be achieved with smaller networks, as only a small portion of the computation is necessary [28].

In conclusion, flying at such altitudes always involves interactions with people. Human−drone interaction (HDI) is a well-developed industry with extensive work covering various topics. Current interdisciplinary research aims to integrate technical and social interaction aspects and develop corresponding UAM systems [28]. For low and medium flight altitudes, recent research has primarily focused on applying deep learning to overcome the challenges of object recognition from the perspective of a UAM-mounted camera. Abundant datasets and global concerns are emerging in this context, indicating that low and medium altitudes are likely to benefit the most from deep learning advancements in the coming years.

Finally, while this work references the integration of various sensors and the addition of 3D information, further research would be necessary to thoroughly investigate UAM object detection using multiple data sources, which is beyond the scope of this chapter. However, the suggested evaluation reveals that sensors such as thermal imaging and 3D cameras are advanced enough to enhance detection performance. Event-based cameras, which provide object identification based on optical flow and motion correction, are still in their early stages but represent a promising research area. In the realm of sensor fusion, the performance of exteroceptive and proprioceptive sensors can be improved. Learning from multimodal sensors presents a unique opportunity to capture correspondences across modalities and develop a comprehensive understanding of the scene. In the years to come, deep neural networks designed specifically for UAM with multimodal input data may emerge.

5.5 Summary

This book chapter provides a wide-ranging analysis of deep learning research and development efforts for autonomous cars, emphasizing UAMs. Thanks to deep learning techniques, UAMs are frequently equipped with cognitive abilities, including situational awareness, decision-making, learning, and adaptation. There are still many unanswered questions about the application of deep learning in UAMs, and there are also many potentials and problems to be resolved. The majority of current research in this field is done using simulated settings and virtual surroundings. This suggests that there are still significant challenges to be overcome when testing urban air mobility (UAM) systems in real-world conditions with external disturbances and noise elements like wind and ground effects. The disparity between controlled virtual environments and real-world settings can introduce uncertainties for reinforcement learning and deep learning methods when applied in actual environments. However, advanced deep learning techniques and adaptive algorithms hold tremendous potential for assisting UAMs in critical applications for the civilian sector. The utilization of UAMs in scenarios involving situational awareness, emergency response, intelligent transportation, smart cities, urban development, and law enforcement represents just a fraction of the potential applications made possible by recent advancements in deep learning modeling and high-performance computing.

References

[1] Walia, K. "VTOL UAV Market 2025 Research Report—Industry Size & Share." (2019).

[2] V Valavanis, Kimon P. "Advances in Unmanned Aerial Vehicles: State of the Art and the Road to Autonomy." (2007).

[3] Tai, Lei, and Ming Liu. "Deep-learning in Mobile Robotics - from Perception to Control Systems: A Survey on Why and Why not." ArXiv abs/1612.07139 (2016): n. page.

[4] Martínez, Carol, et al. "Towards autonomous detection and tracking of electric towers for aerial power line inspection." 2014 International Conference on Unmanned Aircraft Systems (ICUAS) (2014): 284-295.

[5] Olivares-Méndez, Miguel Angel, et al. "Towards an Autonomous Vision-Based Unmanned Aerial System against Wildlife Poachers." Sensors (Basel, Switzerland) 15 (2015): 31362 - 31391.

[6] Carrio, Adrian, et al. "UBRISTES: UAV-Based Building Rehabilitation with Visible and Thermal Infrared Remote Sensing." ROBOT (2015).

[7] Li, Liujun, et al. "Real-time UAV weed scout for selective weed control by adaptive robust control and machine learning algorithm." (2016).

[8] Lu, Yuncheng, et al. "A survey on vision-based UAV navigation." Geospatial Information Science 21 (2018): 21 - 32.

[9] Mozaffari, Mohammad, et al. "Mobile Unmanned Aerial Vehicles (UAVs) for Energy-Efficient Internet of Things Communications." IEEE Transactions on Wireless Communications 16 (2017): 7574-7589.

[10] Commission Directive (EU) 2019/514 of 14 March 2019 Amending Directive 2009/43/EC of the European Parliament and of the Council as Regards the List of Defence-Related Products (Text with EEA Relevance). Available online: https://eur-lex.europa.eu/legal-content/EN/T XT/?uri=uriserv:OJ.L_.2019.089.01.0001.01.ENG(2020).

[11] Siciliano, B et al. "Springer Handbook of Robotics." (2016).

[12] Dalamagkidis, K et al. Classification of UAVs: In Handbook of Unmanned Aerial Vehicles; Springer: Dordrecht, The Netherlands, 2015; pp. 83–91. [CrossRef]

[13] Baca, T et al. "Autonomous landing on a moving car with unmanned aerial vehicle." (2017).

[14] Gade, Rikke, and Thomas BaltzerMoeslund. "Thermal cameras and applications: a survey." Machine Vision and Applications 25 (2013): 245-262.

[15] Portmann, J et al. "People detection and tracking from aerial thermal views." 2014 IEEE International Conference on Robotics and Automation (ICRA) (2014): 1794-1800.

[16] Mitrokhin, Anton, et al. "Event-Based Moving Object Detection and Tracking." 2018 IEEE/RSJ International Conference on Intelligent Robots and Systems (IROS) (2018): 1-9.

[17] Afshar, Saeed, et al. "Event-Based Feature Extraction Using Adaptive Selection Thresholds." Sensors (Basel, Switzerland) 20 (2019): n. page.

[18] Sánchez-López, José Luis, et al. "A Multi-Layered Component-Based Approach for the Development of Aerial Robotic Systems: The Aerostack Framework." Journal of Intelligent & Robotic Systems 88 (2017): 683-709.

[19] Endsley, Mica R. "Toward a Theory of Situation Awareness in Dynamic Systems." Human Factors: The Journal of Human Factors and Ergonomics Society 37 (1995): 32 - 64.

[20] Siegwart, Roland Y. et al. "Introduction to Autonomous Mobile Robots." (2004).

[21] Sánchez-López, José Luis, et al. "A Multi-Layered Component-Based Approach for the Development of Aerial Robotic Systems: The Aerostack Framework." Journal of Intelligent & Robotic Systems 88 (2017): 683-709.

[22] Khamis, AlaaM. "Smart Mobility: Exploring Foundational Technologies and Wider Impacts." Smart Mobility (2021): n. page.

[23] Endsley, Mica R. et al. "Designing for Situation Awareness: An Approach to User-Centered Design." (2003).

[24] Nasa. "NASA Space Technology Roadmaps and Priorities: Restoring NASA's Technological Edge and Paving the Way for a New Era in Space." (2019).

[25] T. M. Mitchell, Machine Learning, vol. 45 (37), McGraw Hill, Burr Ridge, Ill, USA, 1997.

[26] I. Goodfellow, Y. Bengio, and A. Courville, Deep Learning, MIT Press, Cambridge, Mass, USA, 2016.

[27] Hochreiter, Sepp and Jürgen Schmidhuber. "LSTM can Solve Hard Long Time Lag Problems." NIPS (1996).

[28] Cohen, Adam, et al. "Reimagining the Future of Transportation with Personal Flight: Preparing and Planning for Urban Air Mobility." (2020).

[29] Alonso, William. "Location And Land Use." (1964).

[30] Antcliff, Kevin R. et al. "Silicon Valley as an Early Adopter for On-Demand Civil VTOL Operations." (2016).

[31] Balac, Milos, et al. "Towards the integration of aerial transportation in urban settings." (2017).

[32] Balac, Milos et al. "Demand Estimation for Aerial Vehicles in Urban Settings." IEEE Intelligent Transportation Systems Magazine 11 (2019): 105-116.

[33] Banister, David, et al. "Sustainable Cities: Transport, Energy, and Urban Form." Environment and Planning B: Planning and Design 24 (1997): 125 - 143.

[34] Boarnet, Marlon G. "A Broader Context for Land Use and Travel Behavior, and a Research Agenda." Journal of the American Planning Association 77 (2011): 197 - 213.

[35] Bohman, Helena, and Desirée Nilsson. "The impact of regional commuter trains on property values: Price segments and income." Journal of Transport Geography 56 (2016): 102-109.

[36] Brown, Arthur, and Wesley L. Harris. "A Vehicle Design and Optimization Model for On-Demand Aviation." (2018).

[37] Brown, Jay R., and Alfred L. Guiffrida. "Carbon emissions comparison of last-mile delivery versus customer pickup." International Journal of Logistics Research and Applications 17 (2014): 503 - 521.

[38] Burchell, Robert W., and Naveed Akhter Shad. "The Evolution of the Sprawl Debate in the United States." (1999).

[39] Daniel G. Chatman and Nicholas Klein. \Immigrants and Travel Demand in the United States: Implications for Transportation Policy and Future Research." (2009).

[40] Clothier, Reece, et al. "Risk Perception and the Public Acceptance of Drones." Risk Analysis 35 (2015): n. page.

[41] Daziano, Ricardo A. et al. "Are Consumers Willing to Pay to Let Cars Drive for Them? Analzing Response to Autonomous Vehicles." TransportRN: Transportation Modes (2017): n. page.

[42] Graaff, Ad de. "Aviation safety, an introduction." Air & Space Europe 3 (2001): 203-205.

[43] D. N. Dewees. "The Effect of a Subway on Residential Property Values in Toronto." (1976).

[44] Dr. Graham Drozeski. "Vertical Take-off and Landing (VTOL): Emerging and Transformational Capabilities."

[45] Dube, Jean, et al. "Commuter rail accessibility and house values: The case of the Montreal South Shore, Canada, 1992–2009." Transportation Research Part A-policy and Practice 54 (2013): 49-66.

[46] Duranton, Gilles and Matthew Adam Turner. "The Fundamental Law of Highway Congestion: Evidence from the US§." (2008).

[47] "Fast-Forwarding to a Future of On-Demand Urban Air Transportation." (2016).

[48] Filotas, L. T.. "Vortex-induced helicopter blade loads and noise." Journal of Sound and Vibration 27 (1973): 387-398.

[49] Garrow, Laurie A. et al. "Conceptual Models of Demand for Electric Propulsion Aircraft in Intra-Urban and Thin-Haul Markets." (2018).

[50] Garrow, Laurie A. et al. "Forecasting Demand for On-Demand Mobility." (2017).

[51] Garrow, Laurie A. et al. "If You Fly It, Will Commuters Come? A Survey to Model Demand for eVTOL Urban Air Trips." 2018 Aviation Technology, Integration, and Operations Conference (2018): n. page.

[52] Genevieve Giuliano and Susan Hanson. "The Geography of Urban Transportation." (2017).

[53] Hall, Bronwyn H, and Beethika Khan. "Adoption of New Technology." IO: Productivity (2003): n. page.

[54] Hansen, Walter G. "How Accessibility Shapes Land Use." Journal of The American Planning Association 25 (1959): 73-76.

[55] Holmes, Bruce J. "A Vision and Opportunity for Transformation of On-Demand Air Mobility." (2016).

[56] Hu, Yue, et al. "Impacts of Covid-19 mode shift on road traffic." arXiv: Physics and Society (2020): n. page.

[57] Kasliwal, Akshat, et al. "Role of flying cars in sustainable mobility." Nature Communications 10 (2019): n. page.

[58] Kittelson and Incorporated Associates. "Transit Capacity and Quality of Service Manual, 3rd Edition." (2013).

[59] Kloostra, Bradley and Matthew J. Roorda. "Fully autonomous vehicles: analyzing transportation network performance and operating scenarios in the Greater Toronto Area, Canada." Transportation Planning and Technology 42 (2019): 112 - 99.

[60] Lawson, John J. "Flight Path: Taking to the Skies to Solve Congestion." (2007).

[61] Maerivoet, Sven and Bart De Moor. "Transportation Planning and Traffic Flow Models." arXiv: Physics and Society (2005): n. page.

[62] Moore, Mark Douglass. "Misconceptions of Electric Aircraft and their Emerging Aviation Markets." (2014).

[63] Moore, Mark Douglass, et al. "High-Speed Mobility Through On-Demand Aviation." (2013).

[64] Nail, Paul R. et al. "Threat causes liberals to think like conservatives." Journal of Experimental Social Psychology 45 (2009): 901-907.

[65] Newman, Peter. "Sustainable transport for sustainable cities." (2006).

[66] Nneji, Victoria Chibuogu, et al. "Exploring Concepts of Operations for On-Demand Passenger Air Transportation." (2017).

[67] Nneji, Victoria Chibuogu, et al. "Functional Requirements for Remotely Managing Fleets of On-Demand Passenger Airraft." (2018).

[68] Pradeep, Priyank, and Peng Wei. "Energy Efficient Arrival with RTA Constraint for Urban eVTOL Operations." (2018).

[69] Rasouli, Soora and Harry J. P. Timmermans. "Activity-based models of travel demand: promises, progress, and prospects." International Journal of Urban Sciences 18 (2014): 31 - 60.

[70] Reiche, Colleen, et al. "Urban Air Mobility Market Study." (2018).

[71] Rothfeld, Raoul L. et al. "Agent-based Simulation of Urban Air Mobility." 2018 Modeling and Simulation Technologies Conference (2018): n. page.

[72] Rothfeld, Raoul L. et al. "Potential Urban Air Mobility Travel Time Savings: An Exploratory Analysis of Munich, Paris, and San Francisco." Sustainability (2021): n. page.

[73] Senzig, David A. et al. "Sound Exposure Level Duration Adjustments in UAS Rotorcraft Noise Certification Tests." (2018).

[74] Shaheen, Susan A. et al. "The Potential Societal Barriers of Urban Air Mobility (UAM)." (2018).

[75] Sinclair, Robert R.. "VON THNEN AND URBAN SPRAWL." Annals of The Association of American Geographers (1967): n. page.

[76] Stopher, Peter, and John Stanley. "Introduction to Transport Policy: A Public Policy View." (2014).

[77] Thipphavong, David P. et al. "Urban Air Mobility Airspace Integration Concepts and Considerations." 2018 Aviation Technology, Integration, and Operations Conference (2018): n. page.

[78] Uber, Inc. Uber Air Vehicle Mission and Requirements.

[79] NASA Urban Air Mobility. Uam-Market-Study-Executive-Summary

[80] OAR US EPA. Fast Facts on Transportation Greenhouse Gas Emissions.

[81] Vascik, Parker D. and Robert John Hansman. "Evaluation of Key Operational Constraints Affecting On-Demand Mobility for Aviation in the Los Angeles Basin: Ground Infrastructure, Air Traffic Control, and Noise." (2017).

[82] Vascik, Parker D. and Robert John Hansman. "Scaling Constraints for Urban Air Mobility Operations: Air Traffic Control, Ground Infrastructure, and Noise." 2018 Aviation Technology, Integration, and Operations Conference (2018): n. page.

[83] Voom's Helicopter Commuting Service Launches in Mexico City.

[84] Waddell, Paul. "Integrated Land Use and Transportation Planning and Modelling: Addressing Challenges in Research and Practice." Transport Reviews 31 (2011): 209 - 229.

[85] Waddell, Paul, et al. "An Integrated Pipeline Architecture for Modeling Urban Land Use, Travel Demand, and Traffic Assignment." ArXiv abs/1802.09335 (2018): n. page.

[86] Waraich, Rashid A. et al. "Performance improvements for large-scale traffic simulation in MATSim." (2015).

[87] Pavan Yedavalli. "Designing and Simulating Urban Air Mobility Vertiport Networks under Land Use Constraints." Transportation Research Board. (2021).

[88] Yedavalli, Pavan. "An Assessment of Public Perception of Urban Air Mobility (UAM)." (2019).

[89] Dai, Xi et al. "Automatic obstacle avoidance of quadrotor UAV via CNN-based learning." Neurocomputing 402 (2020): 346-358.

[90] Liew, Chun Fui and TakehisaYairi. "Companion Unmanned Aerial Vehicles: A Survey." ArXiv abs/2001.04637 (2020): n. page.

6

Reinforcement Learning for Automated Electric Vertical Takeoff and Landing Decision Making of Drone Taxi

M. Shyamala Devi[1], R. Aruna[2], N. Yuvaraj[3], and Sri Preethaa K. R.[3]

[1]Department of Computer Science and Engineering, Panimalar Engineering College, Chennai, Tamil Nadu, India
[2]VelTech Rangarajan Dr. Sangunthala R&D Institute of Science and Technology, India
[3] School of Computer Science and Engineering, Vellore Institute of Technology, Vellore, India
E-mail: drmshyamaladevi@panimalar.ac.in; drraruna@veltech.edu.in; yuvaraj.n@kpriet.ac.in; k.r.sripreethaa@kpriet.ac.in

Abstract

Unmanned aerial vehicles, in particular drone taxis, have been utilized for crisis management, freight and distribution, wildlife surveillance, catastrophe observation, and passenger transport services. The proliferation of drone taxis in airspace throughout the world will inevitably result in fully autonomous drone taxis. One of the forthcoming upcoming heavy demand transportation systems is the urban aerial mobility system, which includes drone taxi. Electric Vehicle Takeoff and Landing (eVTOL) is one of these urban aerial mobility systems that is used to detect passengers, glide to those areas, board the passengers, and carry the passengers to their respective destinations. By progressively acquiring the optimal control strategy, deep reinforcement learning is a scaffolding paradigm that gets around the limitations of traditional methodologies. Deep reinforcement learning techniques have become increasingly desired in recent years toward electric vertical takeoff and landing since they might contain scalable learning capabilities given the complex drone taxi traffic situations in real life. This chapter provides an analysis of how reinforcement learning could extend its applicability for the electric vertical takeoff and landing for the effective aerial mobility system for the drone taxis toward detection, loading, and shipping of passengers to the respective destinations.

Keywords: Reinforcement learning, drone taxi, aerial mobility, eVTOL

6.1 Introduction

Flexible and dependable network supporting systems are more important, which always have high preference for real-time monitoring services [1]. Unmanned aerial vehicles (UAVs), sometimes known as drones, are now regarded as essential network components for delivering flexible and dependable network services, such as mobile surveillance applications using UAVs. UAVs have exhibited their ability to adapt and dynamically update the locations of the surveillance UAVs by using their mobility feature. For the cooperative task, there needs to be an energy-efficient navigation control for multiple UAVs so that the mission can be done with limited resources and time. Since the amount of energy a UAV uses depends on how long it runs, a UAV's performance is directly linked to how well it uses energy. Control complexity is a typical issue that must be resolved in order to develop an energy-efficient model for multiple UAV controls. When UAVs work together on cooperative missions, the decision of one UAV influences the decision of the other UAVs. In addition, there is an exponential rise in complexity as the number of UAVs grows. UAVs enhance their mobility to provide line-of-sight (LOS) CCTV-based vision-enabled surveillance services, hence guaranteeing the reliability of monitoring services. The on-demand deployment of surveillance UAVs enables them to continuously shift their positions in order to cover a larger region. UAVs are also a good option for strong and flexible mobile CCTV-based surveillance because they are cheap and can be used for many different things. However, UAVs are typically not adequately maintained, which increases their safety risks. For instance, engine shutdowns caused by crashes with other planes or the ground can damage the hardware of the wireless base station. UAVs' onboard power is also used by their mobility and communication support functions. Communication, surveillance, and mobility all use a lot of power; so the amount of power they use needs to be checked on a regular basis. As a result, an autonomous surveillance UAVs management system is needed to make UAV-based network surveillance services more reliable and stable. It is important to both reduce the amount of energy used and improve the reliability of network surveillance services. This is because the movements and deployments of target objects and nearby UAVs can be unpredictable [2].

Drones integrated basic moralities such as infinite mobility, self-governance, and data processing. With the aid of cutting-edge technologies of data collecting, they grant access to locations and enable their analysis [3]. These capacities, which were formerly only available to the military, are now becoming more and more integrated into civil sectors. As a result, drones can be used for a variety of purposes, including passenger transportation and novel types of logistics and monitoring. The latest revolution in transportation will be brought about by on-demand mobility (ODM). The primary purpose of ODM is to transport passengers from their points of origin to their desired destinations. Urban aerial mobility (UAM) is the term for the use of aerial vehicles such as electric vertical takeoff and landing (eVTOL) that are widely studied and commercialized nowadays in order to fulfill this objective and system [4]. The sustainable operation of these electric flying taxi services is just as important as providing a much quicker commute. Air taxis could also be a way for people to get to and from work outside of regular work hours. This would make it easier for people to go to and from big cities for things like sports games or nightlife.

Multiple companies, including Uber, Zephyr Airworks, and Airbus, are preparing to launch their next urban on-demand aviation ride service for residents of densely populated cities. They have discussed eVTOL design concepts and how prototypes will be tested, but city planners are still taking steps to facilitate this new mode of transportation [5]. Thus, forecasting demand for these air taxi services is critical in order for manufacturing organizations, investors, and city planners to be prepared for the opening of this aviation operation. Vertistops and vertiports are two types of physical infrastructure assets suggested for use in air taxi operations. Vertistop is an innovative rooftop helipad that can typically accommodate only one air taxi or other electric vertical takeoff and landing vehicle at a time. These facilities make it possible for air taxis to land and take off quickly in order to load and unload passengers. Alternatively, vertiport is a larger infrastructure that can accommodate multiple eVTOLs simultaneously. At vertiports, there are many structures with landing pads for customers to use. Other air taxi services, such as vehicle inspection and maintenance, charging, and docking, are also done. These stations could be added to larger buildings or used in other places, like highway roundabouts and open parking lots.

The operations of the air taxi system are expected to be connected to public transportation or on-demand taxi services. This will lead to on-demand, multi-leg, multi-modal transportation from door to door. Specifically, the majority of rides consist of three legs or segments. The first part of the trip is getting the customer from his or her actual pickup location to a nearby skyport (i.e., vertiport or vertistop). The primary leg involves the customer being taken by air taxi to their desired skyport. In the third segment, the customer is finally taken to where they need to be dropped off. For the first and last parts of the trip, you could walk, take a regular taxi service like Uber or Lyft, take the subway, or take the bus. Even though air taxis could be a convenient way for people to get around big cities, there are a number of problems that need to be solved before they can be used. For instance, it is crucial to implement a system that makes intelligent real-time dispatching and routing decisions to reduce ride time and costs for customers. Also, the scheduling systems should make sure that flight operations are set up in the best way possible to increase the rate at which demand is met. Estimating how many people want to use air taxi services at different times and in different places is important to deal with the problems listed above and make a number of other important decisions [6].

The purpose of this work is to devise a reinforcement-learning-based method for estimating the electric vertical takeoff and landing for the efficient aerial mobility system for the passengers toward detection, loading, and shipping to the respective destinations. Several ride-related activities such as pickup and drop-off locations, distance, time of day and day of week, and environment-related issues such as temperature, presence of rain or snow, and visibility factors are derived from the dataset and used as classifiers for a reinforcement learning approach.

6.2 Literature Review

In this section, we look at the research on existing and new on-demand transportation services and how to predict demand for air taxi services. A lot of work has been done recently on how to optimize the deployment of UAVs for cellular services. This includes optimization-based coverage control, which is important for surveillance, power allocation reduction, and the deployment of UAVs, taking into account various uncertainties such as users' changing needs, optimal path-planning for multiple UAVs that can provide wireless services to the cell edge individual user on convex relaxation technique, and an analytic model for the optimal deployment of UAVs for cellular services [7].

Even though the past work does an excellent work of achieving their goals, all of the ways to solve the problems are centralized optimization problems. With these methods, it is impossible to get an online (computational) solution for unmanned aerial vehicle (UAV) enabled networks that are highly dynamic and spread out [8].

Approaches that are based on machine learning work well to solve these types of challenges in a distributed environment. In this study, a new multi-agent deep reinforcement learning (DRL) scheme is designed for distributed computation over UAVs, taking into account both users and multiple UAVs. This is to handle the high dynamics of a UAV-enabled network, such as users' increasing demands or neighboring UAVs breaking down [9]. Machine learning techniques have been used to improve the performance of UAV-based computing. For example, an ML-based approach for autonomous trajectory optimization [10] and the optimization of UAV location in a downlink system with a joint K-means and expectation maximization (EM) based on a Gaussian mixture model (GMM) [11], and dynamic optimization of the locations of UAVs in a VLC-enabled UAV-based network for mini-UAVs [12, 13].

Several research works employ DRL approaches in UAV network systems, such as the meta-reinforcement learning based path-planning for UAVs in dynamic and unknown wireless network settings [14], the Q-learning method based dynamic position planning of UAVs in a non-orthogonal multiple access (NOMA) based wireless network [15, 16], and the optimization for UAV optimal energy consumption control taking into account communication coverage, fairness, and connection. The new model of multi-agent deep reinforce learning lets each agent work together to do tasks by direct interaction with other agents and making their own decisions. Compared to a traditional model, this one can be used in situations with multiple agents, such as controlling multiple robots, playing multiplayer games, and controlling multiple UAVs, among other things [17]. A UAV can move in a much wider range than a vehicle on the ground, which moves on a 2D plane. The optimization of unmanned aerial vehicle trajectories is actively discussed in major UAM research results. Optimal UAV trajectories can be calculated using a variety of techniques, including deep reinforcement learning (DRL) [18] and a convex optimization framework [19]. In this case, the DRL approach is appropriate for the UAV mobile network, which operates in real time in an unpredictable environment. Moreover, in the UAV trajectory optimization problem, multiple UAVs must cooperate and coordinate to determine the globally optimal trajectories for each unmanned aerial vehicle.

As a result, multi-agent deep reinforcement learning (MADRL) should be prioritized over other DRL methods [20, 21]. Among them, this chapter considers designing multi-UAV trajectory optimization for passenger transportation while taking QMIX into account [22]. QMIX is a well-known distributed MADRL algorithm, which is why a QMIX-based MADRL algorithm is being thought of for multi-drone taxi trajectory optimization. It is important to note that distributed computation is required because centralized computation cannot handle large numbers of eVTOLs in real time. As a result, QMIX, a distributed MADRL, is used rather than other centralized algorithms, such as communication neural network (CommNet) [23] and game abstraction mechanism based on two-stage attention network (G2ANet) [24].

However, the decision-making of deterministic multi-UAVs in uncertain situations cannot benefit from these broad ML techniques. This is mostly because the above studies do not take into account the UAV-based network system's partially observable multi-agent environment, where each agent has different information and UAVs cannot fully share information [25]. In addition, these methods only focus on the network communication system, even when visual information is available. This shows how much more research needs to be done on how multiple surveillance UAVs can talk to each other. A unique technique is required to share partially observed information from a particular UAV to other UAVs in a volatile and uncertain environment [26].

The sustainable operation of these electric flying taxi services is just as important as providing a much quicker commute. Air taxis could also be a way for people to get to and from work outside of regular work hours. This would make it easier for people to go to and from big cities for things like sports games or nightlife. Multiple companies, including Uber, Zephyr Airworks, and Airbus, are preparing to launch their next urban on-demand aviation ride service for residents of densely populated cities [27]. They have discussed eVTOL design concepts and how prototypes will be tested, but city planners are still taking steps to facilitate this new mode of transportation [28]. Thus, forecasting demand for these air taxi services is critical in order for manufacturing organizations, investors, and city planners to be prepared for the opening of this aviation operation. Over the past few decades, taxis have become one of the most widely used on-demand urban public transportation services in countries all over the world [29]. In recent years, the proliferation of technology, combined with the rapidly increasing population in metropolitan cities, has paved the way for on-demand door-to-door ride-sourcing taxi services such as Uber and Lyft, which provide passengers with both individual and shared/pooled ride options. These mobility services are said to reduce traffic congestion in congested urban areas regardless of "willingness to rideshare" levels [30]. A number of studies have been conducted to provide the framework for taxi companies in dense cities such as New York, Atlanta, and Singapore to begin and successfully implement effective ridesharing services [31]. Several studies have centered on the creation of a framework for estimating taxi demand [32]. For example, to predict passenger demand in 30-minute intervals, the company used streaming data and time series forecasting techniques. In recent years, however, the majority of research has employed machine learning algorithms and used ride-related factors or GPS trace data, such as pickup location, drop-off location, pickup time, and drop-off time, as predictors for predicting taxi passenger demand [33].

Models designed to optimize taxi service dispatch and routing can take advantage of the information provided by the aforementioned prediction models for passenger demand in different locations and times. For real-time matching and sequencing of on-demand ridesharing services, a reinforcement learning approach predicts the constrained optimization solution [34, 35]. Their research showed that on-demand taxi services would improve their service rate, the amount of time passengers had to wait, and the distance each vehicle traveled if a trip delay of 5 minutes or more was allowed. A serviceable algorithm for large-scale vehicle routing in the real world was suggested. Combining the benefits of local search and global optimization, their algorithm generates near-optimal solutions [36]. Even though the above-mentioned on-demand services use road transport, UAM thinks of a new way to do on-demand transportation that could make the above-mentioned services better. In particular, the UAM concept can be broken down into three groups: drones that are mostly used to deliver packages, self-driving air metro services that work like public transportation, and air taxi services that allow people or small groups to move around on demand, like NASA [37].

A lot of research has been done on the use of airborne drones for logistics operations. Several of the issues related to the implementation of delivery drones, including security, privacy, and acceptance, also exist for urban passenger aviation services. Over the past few years, logistics companies all over the world have been testing prototypes for air taxis. The various concept vehicles that have been developed can be divided into three categories: quadrotor, side-by-side, and lift-cruise aircraft [38]. These concept vehicles have undergone pilot and feasibility testing in a number of cities around the world. In the United States, Uber has launched test operations in Dallas and Los Angeles, and in other countries, it intends to expand into markets like Dubai, Tokyo, Singapore, London, and Bangalore. Kitty Hawk, another company, also plans to introduce an air taxi service (called Cora) to the New Zealand market sometime in 2018. Also, Vahana by Airbus, a self-piloted eVTOL, has finished its flight tests in Oregon. Companies like Rolls-Royce, Boeing, and Martin Jetpack are also planning to get into the air taxi business.

6.3 Deep Reinforcement Learning Based eVTOL for Drone Taxi

In deep reinforcement learning [39], agents are modeled as eVTOL vehicles that are capable of computing the optimum route for transporting passengers while taking into account factors like passenger activity, conflicts, and battery health in drone taxi. The drone taxi eVTOL reference model is shown in Figure 6.1.

The machine learning technique known as reinforcement learning has the potential to significantly improve the effectiveness of aerial mobility services by enabling electric vertical takeoff and landing. Sequential decision-making

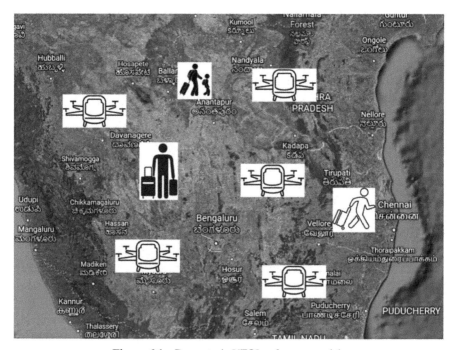

Figure 6.1   Drone taxi eVTOL reference model.

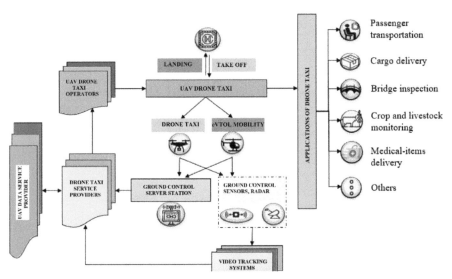

Figure 6.2 Drone taxi eVTOL model applications.

problems are addressed in the field of reinforcement learning, which is distinct from supervised and unsupervised learning. Reinforcement learning [40] learns how to behave in a specific environment in order to increase the reward. Supervised learning is made up of three components: actions, rewards, and observations. In the ecosystem, an agent can take action. The reward acts as the first point of contact between an agent and its environment, providing feedback on whether a recent reaction was successful. Observation is the representation of an environment's state, which consists of information that the agent receives from the context. A method of learning based on prior experience known as reinforcement learning encompasses a variety of circumstances and studies and includes the advantage of not necessitating the compilation of learning data. The applications of drone taxi with eVTOL aerial mobility service is shown in Figure 6.2.

Reinforcement learning is a suitable method for developing an aerial mobility service for electric vertical takeoff and landing in drone taxi applications because it has the advantage of being adaptable in responding to the circumstances in the region based on an optimal policy.

6.4 Deep Q-Learning Network Based eVTOL for Drone Taxi

Figure 6.3 illustrates an action-value function, which is a relationship that may directly compute the optimum action-value function.

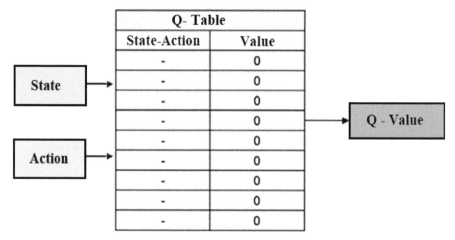

Figure 6.3   Q-learning table.

The Q-function in Q-learning determines whether to do a particular action in a particular state using approximations of the Q-values for each state−activity pair. Utilizing the experience replay technique has the advantage of eliminating the strong association between learning material and experience. As a result, over-fitting is prevented and consistent learning is made possible. By utilizing the target network, it also gets rid of learning uncertainty when using just one network. To establish the precise Q-value, the neural network was partitioned to promote smooth learning. In order to build electric vertical takeoff and landing models, such as the simulation of intelligent passenger detection, landing, and shipping, deep Q-networks [40] are neural networks that use deep Q-learning. The agent then uses a Markov decision process to choose an action from a set of alternatives, changing its environment from the state it is currently in to state $dronestate_{t+1}$, which is the state that will come next. Depending on the previous action, the agent is rewarded, $dronereward_{t+1}$, for completing the transition. The agent's goal is to select the activity that will result in the greatest payout, given the circumstances. In typical deep Q-learning, entries in a table of Q-values Q (droneState, eVTOLagent) are updated when the agent gains new knowledge. Figure 6.4 illustrates the deep Q-network's state and action paradigm. In a deep Q-network learning model, an agent is placed in an environment and instructed to observe its status, or $dronestate_t$.

To determine the ideal Q-function $Q * $ (droneState, eVTOLagent), utilize this table. As the drone state space grows, maintaining a database

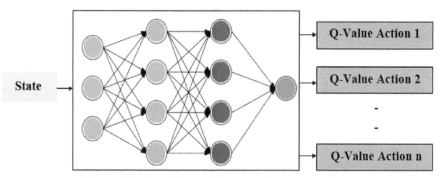

Figure 6.4   Deep Q-network's state and action paradigm.

becomes impractical. To estimate the optimal Q-function value $\mathrm{dronetaxtQ*}$, deep Q-learning network utilizes a deep neural network Q (droneState, eVTOLagent) with eqn (6.1).

$$Q\,(\mathrm{droneState, eVTOLagent}) = \mathrm{dronereward}_{t+1} + \gamma \times \max Q$$
$$* \,(\mathrm{droneState + eVTOLagent + 1})\,.$$
$$(6.1)$$

The agent's goal is to maximize its expected benefit over time. Usually, short-term gains will come before long-term delayed rewards. The system satisfies the Markov chain when the outcome of an operation merely reflects on the previous state and action, as indicated in eqn (6.2).

$$P\left(\begin{array}{c} \mathrm{droneState + 1 \parallel droneState, eVTOLagent,\ droneState - 1,} \\ \mathrm{eVTOLagent - 1, \ldots} \\ (\mathrm{droneState + 1 \parallel droneState, eVTOLagent, dronereward}) \end{array} \right) =$$
$$(6.2)$$

To maximize overall profit, the agent seeks to maximize the return a Q-value. The projected discounted cumulative reward over time, or total yield, can be calculated using eqn (6.3)−(6.4).

$$\mathrm{dronereward}_t = E\,[\mathrm{dronereward}_t + \eth\,\mathrm{dronereward}_{t+1}$$
$$+\eth^2\,\mathrm{dronereward}_{t+2} + \ldots]$$
$$(6.3)$$

$$\mathrm{dronereward}_t = E\left[\sum_{n=0}^{\infty}\eth^n\mathrm{dronereward}_{t+n}\right],$$
$$(6.4)$$

where $\eth$ is the discount factor that has $\eth \in \{0,\,1\}$. It is possible to define two different value functions: a state value function and a state−action value function. With respect to the eVTOL drone vehicle policy, droneState is regarded as eVTOLValue (S). Eqn (6.5)–(6.7) describe the anticipated result of commencing with policy.

$$\text{eVTOLValue (droneState)} = \text{Expectation} \left[\Sigma_{n=0}^{\infty} \eth^n \text{dronereward}_{t+n} \right.$$
$$\left. \| \text{ droneState}_t = \text{droneState} \right]. \tag{6.5}$$

$$\text{eVTOLValue (droneState)} = \text{Expectation} \left[\text{dronereward}_t \right.$$
$$+ \eth \text{ eVTOLValue(droneState}_{t+1})$$
$$\left. \| \text{ droneState}_t = \text{droneState} \right] \tag{6.6}$$

$$\text{eVTOLValue (droneState)} =$$
$$\sum_{n=0}^{\infty} \begin{array}{l} \text{Transition} \left(\text{droneState}, \text{eVTOLValue (droneState)}, \text{droneState}' \right) \\ \left(\text{droneReward} \left(\text{droneState}, \text{eVTOLagent}, \text{ State}' \right) \right. \\ \left. + \eth^n \text{eVTOLValue} \left(\text{droneState}' \right) \right). \end{array}$$
$$\tag{6.7}$$

In a deep Q-network, the Q-values of the eVTOL network system are approximated using deep neural networks. All-purpose coefficient parameter vectors called neural nets can approximate extremely complex and complicated functions. Deep neural network models, which are neural networks, feature a lot of hidden levels. The neural net weights must be upgraded in order to approximate the Q-functions, and this is commonly done using gradient descent and backpropagation algorithms. The deep Q-network equations (6.8) and (6.9) are displayed below.

$$Q_{t+1} \left(\text{dronestate}_t, \text{eVTOLagent}_t \right) = Q_t \left(\text{dronestate}_t, \text{eVTOLagent}_t \right)$$
$$+ \text{learningrate} \times \text{dronereward}_t$$
$$+ \text{TempDifference} \tag{6.8}$$

$$\text{TempDifference} = \eth \text{ maximize } Q_t \left(\text{dronestate}_{t+1}, \text{eVTOLagent}_t \right)$$
$$- Q_t \left(\text{dronestate}_t, \text{eVTOLagent}_t \right). \tag{6.9}$$

In order to take advantage, a cost function is required, which figures out the disparity between the expected and real Q-values of the eVTOL drone taxi deep Q-network. The amount of this error function should indeed be minimized. Since the exact Q-value is unknown, it is possible to approximate it once again using the temporal difference objective of the formula. In addition to future incentives that have been discounted, it displays the overall predicted benefit for all ensuing time steps. Repeated updates to the goal value are then possible. The predicted value of the eVTOL drone taxi network deep Q-network architecture is depicted in Figure 6.5.

The deep Q-network's loss function is created using the squared loss function, as shown in eqn (6.10)−(6.12).

$$\text{eVTOLLoss} = \frac{1}{2}[Q\left(\text{dronestate}_t, \text{eVTOLagent}_t\right)$$
$$- \left(\text{dronereward}_t + \eth\,\text{eVTOLValue}(\,\text{dronestate}_{t+1})\right)]$$
$$(6.10)$$

$$\frac{\partial \text{eVTOLLoss}}{\partial Q(\text{dronestate}_t, \text{eVTOLagent}_t)} = [Q\left(\text{dronestate}_t, \text{eVTOLagent}_t\right)$$
$$- \left(\text{dronereward}_t + \eth\,\text{eVTOLValue}\right.$$
$$\left.(\,\text{dronestate}_{t+1})\right)]$$
$$(6.11)$$

$$Q_{t+1}\left(\text{dronestate}_t, \text{eVTOLagent}_t\right) = [Q_t\left(\text{dronestate}_t, \text{eVTOLagent}_t\right)$$
$$- \left(\text{dronereward}_t + \eth\,\text{eVTOLValue}\right.$$
$$\left.(\,\text{dronestate}_{t+1})\right)].$$
$$(6.12)$$

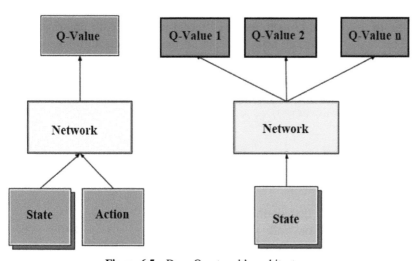

Figure 6.5 Deep Q-network's architecture.

The predicted value of the eVTOL drone taxi network deep Q-network architecture is represented with eqn (6.13).

$$\text{eVTOLLoss}_t(\theta_t) \;=\; E\left[(\text{eVTOLagentValue} \right.$$
$$\left. -Q\left(\text{dronestate}_t, \text{eVTOLagent}_t \colon \theta_t\right)\right)^2\Big]. \qquad (6.13)$$

The neural network is a function of the eVTOL drone taxi deep Q-network from $\text{eVTOLR}^m \to \text{eVTOLR}^n$, and it has an "$n$" dimensional state space and an "m" dimensional action space. Eqn (6.14) gives the target space that the eVTOL drone taxi deep Q-network has identified.

$$\text{eVTOLY}_t^{\text{DQN}} = \text{eVTOLR}_{t+1} + \gamma \underbrace{\max}_{\text{eVTOLagent}} \; \text{Queue}$$
$$\left(\text{eVTOLSignal}_{t+1}, \text{eVTOLagent}; \theta_t^-\right). \qquad (6.14)$$

A multi-layered neural network called the eVTOL drone taxi deep Q-network generates a matrix of action values called eVTOLQ (eVTOLdronestate, θ), where θ denotes the parameters that comprise the network.

6.5 Double Deep Q-Learning Network Based eVTOL for Drone Taxi

In eVTOL drone taxi double deep Q-network [41], two significant functions are learned by periodically updating the two key parameters, resulting in two sets of weights, θ and θ'. For each loop, one group of weights is used to determine the greedy approach, and a different set is used to determine its value. Eqn (6.15) illustrates how to separate the selection and assessment processes in Q-learning and rebuild the position of the object for an easy comparison.

$$\text{eVTOLY}_t^{\text{DQN}} = \text{eVTOLR}_{t+1} \, \gamma \underbrace{\text{Queue}}_{\text{eVTOLagent}} \left(\text{eVTOLSignal}_{t+1}, \right.$$
$$\underbrace{\text{maximum}}_{\text{eVTOLagent}} \text{Queue}\left(\begin{array}{c}\text{eVTOLSignal}_{t+1}, \\ \text{eVTOLagent}, \theta_t^+\end{array}\right); \theta_t^+\right). \qquad (6.15)$$

The eVTOL drone taxi double deep Q-network [42] model's ability to predict with accuracy on both the data it used to train from rather than model, unobserved data is tested using a learning error. An eVTOL drone taxi double deep Q-network model that is overly simple and hence fails to accurately reflect the complexity of the underlying data causes learning mistakes. Despite having more than adequate training data, this can lead to an under-fitting, which occurs when the algorithm fails to recognize important relationships between prominent features and the predicted outputs. Learning error analysis is the procedure to identify, monitor, and evaluate incorrect predictions. The optimum process in the objective is divided into passenger path tracking and action assessment using an eVTOL drone taxi double deep Q-network, and estimation mistakes are reduced with double Q-learning. Despite not being fully decoupled, the optical network in the eVTOL drone taxi double deep Q-network design offers a natural option for the second value function, preventing the need to build additional networks. Eqn (6.16) provides the eVTOL drone taxi double deep Q-network learning error:

$$eVTOLY_t^{DQN} = eVTOLR_{t+1} + \gamma \underbrace{Queue}_{eVTOLagent} (Signal_{t+1},$$

$$\underbrace{maximum}_{eVTOLagent} Queue \left(\begin{array}{c} eVTOLSignal_{t+1}, \\ eVTOLagent, \theta_t^+ \end{array} \right); \theta_t'). \quad (6.16)$$

The eVTOL drone taxi double deep Q-network evaluates the greedy strategy using the traffic network and determines its value using the target network as a result. The eVTOL drone taxi double deep Q-network combines with double Q-learning and deep Q-network. Eqn (6.17) calculates the passenger path target location.

$$eVTOLY_t^{DQN} = eVTOLR_{t+1} + \gamma \underbrace{Queue}_{eVTOLagent} (eVTOLSignal_{t+1},$$

$$\underbrace{maximum}_{eVTOLagent} Queue \left(\begin{array}{c} eVTOLS_{t+1}, \\ eVTOLa, \theta_t^+ \end{array} \right); \theta_t^-). \quad (6.17)$$

The eVTOL drone taxi double deep Q-network model has been used to train deep Q-network and double deep Q-network [43, 44], and the average estimate of the model is calculated. To be more precise, eqn (6.18) represents how full assessment stages of duration for T cycles are employed to decide the estimated average value predictions on a regular basis throughout learning.

$$Average\ Estimate = \frac{1}{T} \sum_{t=1}^{T} \underbrace{maximum}_{eVTOLagent} Queue$$

$$(eVTOLSignal_t, eVTOLagent; \theta_t). \quad (6.18)$$

The dual networks, target network, double Q-network, and prioritized memory relay are all combined to create the double deep Q-network in the eVTOL drone taxi double deep Q-network model system. In the systems, the Q-value is estimated using the current state number and the edge of each activity relative to the other activities. The quantity of a phase eVTOLV (eVTOLScale; θ) represents the expected overall impacts of using stochastic judgments in the following steps. The term eVTOLV (eVTOLScale, eVTOLagent; θ) denotes the benefits that are connected to each activity. The scaling factor, which is determined by the following formula, is created by adding the amount Q and the advantage factor eVTOLVagent. The quantity of a phase eVTOLV (eVTOLScale, eVTOLagent; θ) represents the expected overall impacts of using stochastic judgments in the following steps. The formula eVTOLAgent(eVTOLScale, eVTOLagent;) denotes the benefits that are connected to each activity. The scaling factor is derived from eqn (6.19) by adding the advantage factor eVTOLagent along with the amount eVTOLQ.

$$
\begin{aligned}
\text{Queue}\,(\text{eVTOLSignal}, \text{eVTOLagent}; \theta) &= \text{eVTOL}\ V\,(\text{eVTOL}S; \theta) \\
&+ \Big\{ A\,(\text{eVTOL}S, \text{eVTOLagent}; \theta) - \tfrac{1}{\|\text{eVTOLA}\|} \\
&\sum\nolimits_{\text{eVTOLagent}} \text{eVTOL}A\,(\text{eVTOL}S, \text{eVTOLagent}; \theta) \Big\}.
\end{aligned}
\tag{6.19}
$$

Eqn (6.20) evaluates the mean square error of the target eVTOL drone taxi double deep Q-network performance.

$$
\text{eVTOLMSE} = \sum\nolimits_{a} P\,(S)\,[\text{Queue}_{\text{target}}\,(\text{eVTOLSignal}, \text{eVTOLagent}) - \text{Queue}((\text{eVTOL}S, \text{eVTOLagent}; \theta))]^2.
\tag{6.20}
$$

In order to improve the suggested system's dependability, the eVTOL drone taxi double deep Q-network target is developed using delayed upgrades and experiences replaying methodologies. The eVTOL drone taxi double deep Q-network system learns a dependable and efficient method that significantly reduces the length of the line and the amount of time that eVTOL vehicles must queue in it by employing multiple time-varying passenger path conditions. The eVTOL drone taxi double deep Q-network as shown in Figure 6.6 outperformed traditional drone taxi aerial mobility control systems.

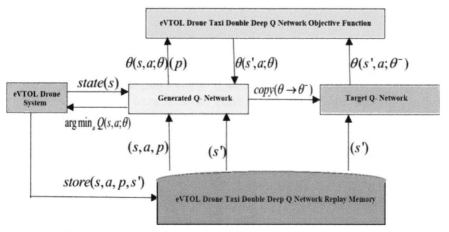

Figure 6.6   eVTOL drone taxi double deep Q-network architecture.

6.6 Multi-agent Deep Q-Learning Network Based eVTOL for Drone Taxi

The reinforcement learning algorithm saves data in the replay buffer before undertaking learning by randomly choosing knowledge in order to remove functional dependencies. However, not all of the data stored in the replay buffer is equivalent. Exceptions might occur depending on the situation; however, a certain experience might be more valuable. Numerous studies have been conducted to rank or describe the importance of learning knowledge in consideration of these characteristics. Nevertheless, the prioritized experience replay method has proven to be the most effective. Multi-agent systems [45] have shown to be a useful tool for creating resolutions to problems that arise in geographical locations. In a distributed architecture, the data, the control operations, or perhaps even both could be spread. When discussing situations where it would be difficult or even impossible for a single creature to understand everything there is to know about the state of a system, the concept of an agent might be helpful. Multi-agent systems [46] could be used in a variety of distributed applications, such as traffic coordination, route load balancing difficulties, and traffic negotiation between drone taxi infrastructure and passengers. It is difficult to establish a multi-agent deep Q-network reinforcement learning that considers both the stability of the eVTOL drone taxi agent's learning dynamics and its capacity to adjust to the other eVTOL drone taxi agents' shifting behavior. For static activities, the objectives frequently define the rules in terms of drone state and drone rewards. Numerous investigations and initiatives are being made to improve learning performance in different applications. It is advised to consider a number of characteristics that have a direct impact on the multi-agent eVTOL drone taxi deep Q-network's performance.

In contrast, a reward-free environment shows that there is a minimal probability of obtaining a clearly defined drone reward in a non-periodic and exceptional setting. A high likelihood of accomplishing a goal, through even unlearned acts, is indicated by an environment where rewards do appear frequently. Eqn (6.21) and (6.22) illustrate how the Q-values for the eVTOL drone taxi multi deep Q-network are determined.

$$Q_t \left(\text{dronestate}_t, \text{eVTOLagent}_t\right) = Q_t \left(\text{dronestate}_t, \text{eVTOLagent}_t\right)$$
$$+ \text{ droneRewardvalue.} \qquad (6.21)$$

$$\text{droneRewardvalue} = \eth \left(\begin{array}{c} \text{dronereward}_{t+1} + \gamma \max Q_t \\ (\text{eVTOLstate}_{t+1}, \text{eVTOLAction}_t) \end{array} \right)$$
$$- Q_t \left(\text{dronestate}_t, \text{eVTOLagent}_t\right). \qquad (6.22)$$

Figure 6.7 depicts the multi-agent-based architecture for the drone taxi applications for the aerial mobility system.

Since it employs a large number of actors in accordance with the unique characteristics of the design, the eVTOL drone taxi multi-agent deep Q-network experience accumulation rate is larger than that of the general DQN.

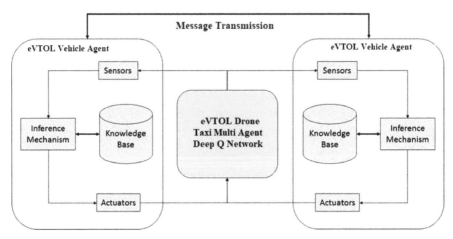

Figure 6.7   Multi-agent-based architecture for the drone taxi.

The method for determining experience value through the deployment of prioritized experience replay based on this architecture is therefore one factor that greatly determines learning performance. The significance of a situation is determined by how much may be learned from it based on experience. These numerical measures can be calculated by comparing the target value also known as the target error and the actual values with eqn (6.23) and (6.24).

$$\text{eVTOLTarget}_t = ((\text{droneReward}_t) + \text{droneRewardTarget}_t - Q_t (\text{dronestate}_{t-1}, \text{eVTOLagent}_{t-1})_t) \tag{6.23}$$

$$\text{droneRewardTarget}_t = \gamma \, Q_{\text{eVTOLtarget}} (\text{dronestate}_t, \max Q_t (\text{dronestate}_t, \text{eVTOLagent}_t))_t) \, . \tag{6.24}$$

Continuous operations and state−action compensate parameterized actions in reinforcement learning. The Q-feature is used to estimate the Q scaling factor value, which is then stored in a database as part of traditional Q-learning. The problem arises because the database calculation costs increase when the action and state spaces are expressed as a Q table. Figure 6.8 depicts the eVTOL drone taxi multi-agent-based reinforcement learning architecture that can have any number of agents as per the requirement.

In contrast, eVTOL drone taxi multi-agent deep Q-network uses a deep neural network to compute the Q-function and determine the Q-value instead of using the Q-table repository. The eVTOL drone taxi multi-agent deep Q-network may be employed to detect problems with large state and action spaces when artificial neural network function cannot. The eVTOL drone taxi multi-agent deep Q-network tackles the problem of inadequate learning

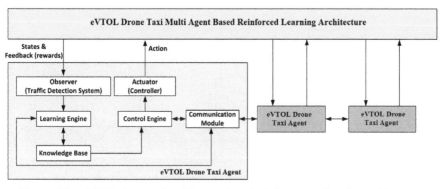

Figure 6.8 eVTOL drone taxi multi-agent-based reinforcement learning architecture.

in Q-learning dependent on a function and an artificial neural network by merging experience replay and target network. Experience replay is a method where an eVTOL drone taxi agent interacts with the environment to gather an incident, but rather than immediately applying this knowledge for learning, saves a portion of the knowledge in storage and subsequently extracts the pattern. The eVTOL drone taxi agents in the eVTOL drone taxi multi-agent framework can communicate with each other.

The aggregate actions of various agents, each of which acts in line with its local knowledge base, enable the eVTOL drone taxi multi-agent network objective to be achieved. Coordination methods can also be established using agent interaction. Each control device agent reacts to the circumstances of the system by performing specific activities. A knowledge base is made available to the eVTOL drone taxi agent to provide it memory and intelligence for decision-making. In addition, an eVTOL drone taxi agent's knowledge base is continuously updated through a series of activity trails. The eVTOL drone taxi agent executes the subsequent action decisions for an actuator depending on the level of knowledge and control signals of the eVTOL drone taxi controller. In discrete time, the eVTOL drone taxi multi-agent control system makes decisions based on the observed aerial mobility conditions of the passengers. The architecture of the eVTOL drone taxi multi-agent deep Q-network is shown in Figure 6.9.

This provides a framework for dealing with difficult situations where a higher degree of control and slightly heightened operations are required. The eVTOL drone taxi multi-agent deep Q-network approach allows training over such action spaces. Furthermore, by interpreting all action-parameters as a single joint input, it compromises the theoretical foundations of the multi-agent deep Q-network.

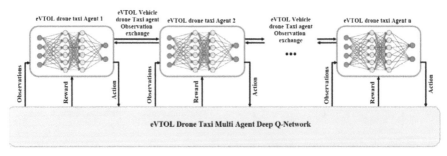

Figure 6.9   eVTOL drone taxi multi-agent deep Q-network architecture.

6.7 Summary

The restriction of flying under 400 feet and avoiding airports to easily integrate UAVs into the airspace is achieved by means of drone taxi. The likelihood of a drone colliding with an airliner will be reduced by 95% or more just by these two factors. However, as a passenger drone is an aircraft, it must fly in the upper airspace. The passengers are transported within cities using drone taxis. Drone taxis are well equipped with contemporary, stealthy, secure, and ecologically friendly features. Lowered heavy traffic, durability, excellent transportation efficiency, and noise reduction are the benefits of drone taxi. The world is moving toward implementing the drone taxi as equivalent to the road transportation as it travels with the shortest route to carry, ship, and land the passengers. Since drone taxi is flying in the airspace, the action coordination is essential for the proper functioning of the drone taxi. The drone taxi is forecasted to grow to 860 million in 2023 that provides intracity transportation. The eVTOLs are intended to move substantial amounts of freight and cargo between airports, manufacturing facilities, distribution centers, and end-users or consumers. A network of supply chains is made up of five main players: producers, processors, wholesalers, merchants, and buyers or consumers. The period of time it takes for an entity to get the products after placing an order is referred to as the lead time. Customers definitely want shorter lead times for deliveries. Additionally, owing to the time restrictions of clients and the growing interest in e-commerce, lead time has grown to be a significant quantity of interest. An integrated supply chain network's decreased lead time suggests the adoption of speedier delivery techniques. Increasing carbon dioxide emissions could be reduced by transitioning from conventional fossil fuel-powered ground vehicles to eVTOLs. Duration and money savings are crucial factors to consider before investing in a project in this cutthroat environment. This chapter attempts to provide the architecture and reference model for the drone taxi extended with eVTOL aerial mobility using reinforcement learning. The concepts behind the model design of drone taxi use the deep reinforcement learning, deep Q-network, double deep Q-network, and multi-agent deep Q-network. This chapter has a number of fascinating extensions that will result in interesting future work. Furthermore, future work of this paper focuses on drone taxi inspections of tunneling, pipelines, antennas, bridges, and other construction structures. Furthermore, future work can look into how drone taxi affects society, including how it might enhance air traffic congestion, delayed flights, and excessive noise levels, as well as how it might affect land vehicle drivers by obstructing their vision and invading their privacy.

References

[1] J. Alonso-Mora, S. Samaranayake, A. Wallar, E. Frazzoli, D. Rus, D, 'On-demand highcapacity ride-sharing via dynamic trip-vehicle assignment', Proc. In National Academy of Sciences, 114(3), pp. 462-467, 2017.

[2] G. Biau, B. Cadre, L. Rouviere, 'Accelerated gradient boosting', Machine Learning, 108(6), pp. 971-992, 2019.

[3] K. M. Gurumurthy, K. M. Kockelman, 'Analyzing the dynamic ride-sharing potential for shared autonomous vehicle fleets using cellphone data from Orlando, Florida', Computers, Environment and Urban Systems, 71, pp. 177-185, 2018.

[4] W. Johnson, C. Silva, 'Observations from Exploration of VTOL Urban Air Mobility Designs', White Paper,

[5] T. Kim, S. Sharda, X. Zhou, R. M. Pendyala, 'A stepwise interpretable machine learning framework using linear regression (LR) and long short-term memory (LSTM): Citywide demand-side prediction of yellow taxi and for-hire vehicle (FHV) service', Transportation Research Part C: Emerging Technologies, 120, pp. 102786, 2020.

[6] S. Liao, L. Zhou, X. Di, B. Yuan, J. Xiong, 'Large-scale short-term urban taxi demand forecasting using deep learning', Proc. In 23rd Asia and South Pacific Design Automation Conference, pp. 428-433, 2020.

[7] Y. Liu, Z. Liu, C. Lyu, J. Ye, 'Attention-based deep ensemble net for large-scale online taxi-hailing demand prediction', IEEE Transactions on Intelligent Transportation Systems, 2019.

[8] Z. Liu, H. Chen, Y. Li, Y, Q. Zhang, 'Taxi demand prediction based on a combination forecasting model in hotspots', Journal of Advanced Transportation, 2020.

[9] M. Lokhandwala, H. Cai, 'Dynamic ride sharing using traditional taxis and shared autonomous taxis: A case study of NYC', Transportation Research Part C: Emerging Technologies, 97, pp. 45-60, 2018.

[10]]J. Kim, J. H. Kim, 'Joint message-passing and convex optimization framework for energy-efficient Surveillance UAV scheduling', Electronics, 9, pp. 1475, 2020.

[11] D. Kwon, J. Jeon, S. Park, J. Kim, S. Cho, 'Multiagent DDPG-based deep learning for smart ocean federated learning IoT networks', IEEE Internet of Things Journal, 7, pp. 9895–9903, 2020.

[12] M. Brittain, P. Wei P, 'Autonomous separation assurance in an high density en route sector: A deep multi-agent reinforcement learning approach', Proc. In IEEE Intelligent Transportation Systems Conference, 2019.

[13] T. Rashid, M. Samvelyan, C. Schroder de Witt, G. Farquhar, J. N. Foerster, S. Whiteson, 'QMIX: Monotonic value function factorization for deep multi-agent reinforcement learning', Proc. In ICML, 2018.

[14] M. Shin, D. Choi, J. Kim, 'Cooperative management for PV/ESSenabled electric vehicle charging stations: A multiagent deep reinforcement learning approach', IEEE Transactions on Industrial Informatics, 16, pp. 3493–3503, 2020.

[15] Y. Liu, W. Wang, Y. Hu, J. Hao, X. Chen, Y. Gao, 'Multi-agent game abstraction via graph attention neural network', Proc. In AAAI, 2020.

[16] X. Yang, P. Wei, 'Scalable multi-agent computational guidance with separation assurance for autonomous urban air mobility operations', AIAA Journal of Guidance, Control, and Dynamics, 43, pp. 1473-1486, 2020.

[17] H. Matsumoto, K. Domae, 'The effects of new international airports and air-freight integrator's hubs on the mobility of cities in urban hierarchies: A case study in East and Southeast Asia', Journal of Air Transport Management, 71, pp. 160-166, 2018.

[18] S. Rajendran, J. Shulman, 'Study of Emerging Air Taxi Network Operation using Discrete-Event Systems Simulation Approach', Journal of Air Transport Management, 2020.

[19] S. Rajendran, E. Pagel, 'Recommendations for emerging air taxi network operations based on online review analysis of helicopter services', Heliyon, 6(12), pp. e05581, 2020.

[20] S. Srinivas, Rajendran, 'A Data-Driven Approach for Multiobjective Loan Portfolio Optimization Using Machine-Learning Algorithms and Mathematical Programming', Proc. In Big Data Analytics Using Multiple Criteria Decision-Making Models, pp. 191-226, CRC Press, 2017.

[21] S. Rajendran, J. Zack, 'Insights on strategic air taxi network infrastructure locations using an iterative constrained clustering approach', Transportation Research Logistics and Transportation Review, 128, pp. 470-505, 2019.

[22] D. Kim, S. Park, J. Kim, J. Y. Bang, S. Jung, 'Stabilized adaptive sampling control for reliable real-time learning-based surveillance systems', Journal of Communications and Networks, 23(2), pp. 129–137, 2021.

[23] H. Huang, A. V. Savkin, 'An algorithm of reactive collision free 3-D deployment of networked unmanned aerial vehicles for surveillanceand monitoring', IEEE Transactions on Industrial Informatics, 16(1), pp. 132–140, 2020.

[24] R. Nawaratne, D. Alahakoon, D. D. Silva, X. Yu, 'Spatiotemporal anomaly detection using deep learning for real-time video surveillance', IEEE Transactions on Industrial Informatics, 16(1), pp. 393-402, 2020.

[25] S. Jung, W. J. Yun, M. Shin, J. Kim, J. H. Kim, 'Orchestrated scheduling and multi-agent deep reinforcement learning for cloud assisted multi-UAV charging systems', IEEE Transactions on Vehicular Technology, 70(6), pp. 5362–5377, 2021.

[26] K. Muhammad, T. Hussain, J. D. Ser, V. Palade, V. H. C. De Albuquerque, 'DeepReS: A deep learning-based video summarization strategy for resource-constrained industrial surveillance scenarios', IEEE Transactions on Industrial Informatics, 16(9). pp. 5938–5947, 2020.

[27] M. Shin, J. Kim, M. Levorato, 'Auction-based charging scheduling with deep learning framework for multi-drone networks', IEEE Transactions on Vehicular Technology, 68(5), pp. 4235–4248, 2019.

[28] Z. Zhang, Y. Xiao, Z. Ma, M. Xiao, Z. Ding, X. Lei, G. K, Karagiannidis, P. Fan, '6G wireless networks: Vision, requirements, architecture, and key technologies', IEEE Vehicular Technology Magazine, 14(3), pp. 28–41, 2019.

[29] S. Park, W.-Y. Shin, M. Choi, J. Kim, 'Joint mobile charging and coverage-time extension for unmanned aerial vehicles', IEEE Access, 9, pp. 94 053–94 063, 2021.

[30] T. T. Nguyen, N. D. Nguyen, S. Nahavandi, 'Deep reinforcement learning for multiagent systems: A review of challenges, solutions, and applications', IEEE Transactions on Cybernetics, 50(9), pp.3826–3839, 2020.

[31] M. Shin, D. Choi, J. Kim, 'Cooperative management for PV/ESS enabled electric vehicle charging stations: A multiagent deep reinforcement learning approach', IEEE Transactions on Industrial Informatics, 16(5), pp. 3493–3503, 2020.

[32] W. J. Yun, S. Jung, J. Kim, J.H. Kim, 'Distributed deep reinforcement learning for autonomous aerial eVTOL mobility in drone taxi applications', ICT Express, 7(1), pp. 1–4, 2021.

[33] S. Jung, W. J. Yun, J. Kim, J. H. Kim, 'Infrastructure-assisted cooperative multi-UAV deep reinforcement energy trading learning for bigdata processing', Proc. In IEEE International Conference on Information Networking (ICOIN), Jeju Island, Republic of Korea, 2021.

[34] Samsung Electronics, 'Samsung brings advanced ultra-fine pixel technologies to new mobile image sensors', 2021.

[35] L. Lu, Y. Hu, Y. Zhang, G. Jia, J. Nie, M. Shikh-Bahaei, 'Machine learning for predictive deployment of UAVs with multiple access', Proc. In IEEE GLOBECOM Workshops, Taipei, Taiwan, 2020.

[36] S. R. Winter, S. Rice, T. L. Lamb, 'A prediction model of Consumer' willingness to fly in autonomous air taxis', Journal of Air Transport Management, 89, pp. 101926, 2020.

[37] National Aeronautics and Space Administration (NASA), Crown Consulting, McKinsey and Company, Ascension Global, Georgia Tech Aerospace Systems Design Lab, 2018a. In: Urban Air Mobility (UAM) Market Study. NASA, White paper,

[38] R. Merkert, J. Bushell, 'Managing the drone revolution: A systematic literature review into the current use of airborne drones and future strategic directions for their effective control', Journal of Air Transport Management, 89, pp. 101929, 2002.

[39] National Aeronautics and Space Administration (NASA), 2018b. In: Observations from Exploration ofVTOL Urban Air Mobility Designs. NASA, White paper,

[40] V. Mnih, K. Kavukcuoglu, D. Silver, A. A. Rusu, J. Veness, M. G. Bellemare, A. Graves, M. Riedmiller, A. K. Fidjeland, G. Ostrovski, 'Human-level control through deep reinforcement learning', Nature, 518(7540), pp. 529–533, 2015.

[41] H. Van Hasselt, A. Guez, D. Silver, 'Deep reinforcement learning with double q-learning', Proc. In Thirtieth AAAI Conference on Artificial Intelligence, pp. 2094–2100, 2016.

[42] H. Van Hasselt, 'Double Q-learning. Advances in Neural Information Processing Systems, 23, pp. 2613–2621, 2010.

[43] L. Zou, M. S. Munir, S. S. Hassan, Y. K. Tun, L. X. Nguyen and C. S. Hong, 'Imbalance Cost-Aware Energy Scheduling for Prosumers Towards UAM Charging: A Matching and Multi-Agent DRL Approach', IEEE Transactions on Vehicular Technology, vol. 73, no. 3, pp. 3404-3420, March 2024, doi: 10.1109/TVT.2023.3328266.

[44] Qiming Zheng, Hongfeng Xu, Jingyun Chen, Dong Zhang, Kun Zhang, Guolei Tang, 'Double Deep Q-Network with Dynamic Bootstrapping

for Real-Time Isolated Signal Control: A Traffic Engineering Perspective', Appl. Sci. 12, 8641, 2022.

[45] Keecheon Kim, 'Multi-Agent Deep Q Network to Enhance the Reinforcement Learning for Delayed Reward System', Appl. Sci. 12, 3520, 2022.

[46] J. Craig, D. Bester, D. Steven James, D. George Konidaris, 'Multi-Pass Q-Networks for Deep Reinforcement Learning with Parameterised Action Spaces', 2019. https://doi.org/10.48550/arXiv.1905.04388.

7

Urban Aerial Mobility Concepts, Modeling, and Challenges: A Review

Premkumar Duraisamy[1], Yuvaraj Natarajan[2], and Sri Preethaa K. R.[2]

[1]Department of Computer Science and Engineering,
KPR Institute of Engineering and Technology, India
[2]School of Computer Science and Engineering, Vellore Institute
of Technology, Vellore, India
E-mail: technicalprem@gmail.com; yuvaraj.n@vit.ac.in;
sripreethaa.kr@vit.ac.in

Abstract

Industry and academia have contributed resources in recent years to create fresh ideas to enhance the performance of urban transportation. Providing traffic congestion-free air transportation services within urban areas is the primary design goal for intelligent urban air mobility (UAM) systems. Among the regular services, specific scenarios need emergency attention, such as medical evacuations, rescue missions, humanitarian missions, intelligence collecting, country current evaluations, weather monitoring, cash distribution, and personnel transfers. One of the focused designs is the beginning of eVTOL (electric vertical takeoff and landing), which is considerably less expensive than helicopters and can be utilized for inspections, valuables transportation, and people's movement. Agent-based modeling is a particularly intriguing technique that may be used to research the interplay between urban systems and transport. The chapter discusses the significance of UAM, ideology, modeling methods, and implementation challenges.

Keywords: Urban systems, urban modeling, vertical ports, agents, personal aviation vehicles.

7.1 Introduction

Researchers, industries, and governments are looking for innovative and applicable solutions to the long travel time of people and goods overland, especially in densely populated areas [1]. With the ever-increasing interest in global consolidation of UAM operations, employing new technologies, and implementing improved types of aircraft, changes in airspace operations and management are in demand. UAM refers to systems that provide highly automated, demand-driven air traffic services for passenger and freight transportation [2]. VTOLs, which fly vertically and land in small areas commonly referred to as vertical ports, are the most prevalent type of vehicle in the UAM environment. Two significant advantages of air transport over ground transport are the higher vehicular speed and the possibility of more direct routes. Considering door-to-door shifts, these benefits reduce passenger travel time using stream transportation systems [3].

The UAM idea of using vertical takeoff and landing aircraft (VTOL) is gaining increasing interest in industry and research. Helicopter-based taxi services exist and continue to live worldwide in major cities, but air mobility with VTOL vehicles will be a new era [4]. However, technological advances enabled the development of the next generation of so-called personal air vehicle (PAV) or electric VTOL vehicles (eVTOLs). The UAM air transport system architecture is depicted in Figure 7.1.

UAM air transport is believed to have less environmental impact than conventional aircraft and other standard transportation modes due to eliminating greenhouse gases, thus reducing air and noise pollution (Bian et al., 2021). As UAM is currently in the planning and testing stages, its related procedures and standards are evolving by legal bodies, aviation authorities, and others. UAM is a new aviation industry with innovative ideas for air transport, so the certification and ownership of UAM operators, regulations, permits, operations, infrastructure, maintenance, and certification procedures are still controversial. Since UAM is a branch with cutting-edge applications and high-tech design concepts, operational quality and flight safety by applicable regulations must be devised by the aviation authorities [5].

Next-generation VTOLs, based on improved electric and distributed propulsion and increased battery energy density, produce less noise and emissions than traditional helicopters, making urban air transport economically viable, and it can be made socially acceptable.

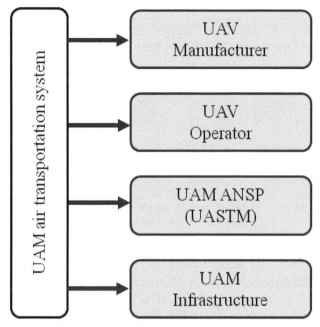

Figure 7.1 UAM air transportation system.

7.2 UAM History

Urban aviation is a familiar idea. The innovators began developing his "flying vehicle" theory in the early 1900s, and by the mid-20th century, early operators were offering regular helicopter rides [6]. This section explores UAM's past, present, and prospects in six phases, as depicted in Figure 7.2.

7.2.1 Phase 1: Conceptual flying car

The UAM concept dates back to the Autoplane, a functional "flying car" developed by Glenn Curtiss around 1917.

7.2.2 Phase 2: UAM as scheduled helicopter services

From the 1950s to the 1980s, many airlines in Los Angeles, New York City, and San Francisco (SF Bay Area) began offering early UAM services using helicopters. In the mid-1950s, New York Airlines provided passenger service between Manhattan and LaGuardia. These early passenger helicopter services in the United States were often made possible by helicopter subsidies

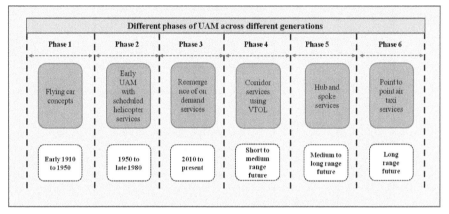

Figure 7.2   UAM phase history.

(withdrawn in 1966) and airmail revenues. From 1965 to 1968 (resumed in 1977), Pan Am operated between Midtown and Worldport at JFK on an hourly basis, with passengers' arrival time of 40 minutes before the plane's departure. Attractive travel offers at JFK, free helicopter-to-air transfers for international business travelers, include "Buy One; He Gets One Free." Service was interrupted in 1977 after a roof crash due to metal fatigue in the landing gear, killing 5 people, 4 on the roof, 1 in the basement, and 59 on the stories. Helicopter services slowly made a comeback in Manhattan in the 1980s. Trump Airlines operated scheduled flights between Wall Street and LaGuardia using Sikorsky S-61 helicopters attached to Trump Shuttle flights. The service was discontinued in the early 1990s when US Airways purchased the Trump shuttle [7].

7.2.3 Phase 3: Re-emergence of on-demand services

Below are recent breakthroughs in unmanned personnel movement and goods delivery, and it is increasing. UAS applications are currently deployed across various domains, including consumer goods delivery, medical samples and emergency supplies transportation, and mapping and surveillance. In recent years, small drones have delivered consumer goods and emergency supplies, which have received extensive industrial attention [8].

A medical facility utilized its UAS as an international service to deliver medicines, emergency medical supplies, laboratory samples, and vaccines. Zipline International uses UAS to distribute blood, vaccines, and drugs in Ghana and Rwanda [9]. This UAS shipping service for testing medical samples was introduced in Switzerland by Matternet and Swiss Post. Some of the remarkable examples of urgent deliveries tested are:

1. Swoop Aero
2. MADRONA project in Belgium
3. Partnership with DHL and Wingcopter in Tanzania (Vanuatu).

To facilitate the integration of safe drones in the United States, the FAA's UAS Integration Pilot Program (IPP) brought together state, local, and tribal agencies and private companies such as UAS operators and manufacturers. A team of nine senior participants investigates different operational approaches that include:

• Parcel delivery (both consumer goods and medical supplies)
• Flying over people or beyond the pilot's sight
• Night operation
• Detection/prevention technology
• Reliability and security of the data link between the pilot and the aircraft

These demonstrations produced numerous use cases for deployment in various formats. For example, Matternet and UPS jointly ship medical supplies to hospitals in North Carolina. Additionally, in April 2021, UPS announced purchasing 10 of its eVTOL aircraft from Beta Technologies, which can carry shipping containers [10]. With the option to buy an additional 150 aircraft, UPS will test the aircraft as part of its high-speed air transportation network. Deloitte and Rady, Children's Institute for Genomic Medicine, is exploring options for shipping test samples through its UAS in California. Royal Mail has launched a COVID experiment in the UK between the mainland and the Isles of Scilly, studying the use of drones to transport parcels.

7.2.4 Phase 4: Corridor services using VTOL

Passenger operations are expected to begin in the later stages of UAM development (Phase 4) using VTOL aircraft. These operations may evolve from Phase 3 services and take the form of regular "air shuttle" services (BLADE, Voom, etc.) along specific flight routes (e.g., between airports and downtown). Based on pre-pandemic market assumptions, some market research agencies predict the potential for large-scale operations and superior service in the late 2020s and early 2030s. Many service providers plan to launch services in the early to mid-2020s. The companies that have a vision of introducing on-demand electric VTOL (eVTOL) flight services over the next decade, to name a few, are EHang in Linz, Austria, aerospace vertical in London in 2022, Lilium in Munich, Orlando, and other cities worldwide by 2024−25. Despite the recovery from the global pandemic, the availability of certified eVTOL aircraft and other conditions could impact previously anticipated deadlines. Companies with unknown release dates include Archer, Joby Aviation (formerly Uber Elevate), Wisk (formerly Kitty Hawk), and Skyrise. Up to 30 eVTOL aircraft will be deployed on the BLADE platform, all owned, operated, and maintained by Wisk. Skyrise plans to introduce autonomous helicopter services as an alternative to VTOL [11].

7.2.5 Phase 5: Hub and spoke services

Phase 5 could see more lavish infrastructure spending to assist "air metro services" as flight expenses come down, and adoption may become more widespread. Multiple flights in step with day among the same hub (or more fabulous touchdown pads that can ever take delivery of more than one plane) and several vertiports (available touchdown pads that can accommodate three planes) would possibly make up those services, which could shape a hub and spoke community over a city region. Vertihubs, for instance, are probably located in busy industrial areas, and vertiports in much less populated residential areas. There are alternatives for scheduling flights at regular durations for the day and sharing flights with extra humans to decrease the fee in step with the seat. However, directional tour styles that might result in more charges and fell load elements due to deadheading may also be inspired by land use, demand, and tour habits (i.e., a plane without passengers or items repositioning to another location). Technological improvements, which include dynamic pricing, state-of-the-art algorithms to cast off deadheading, extended battery capacity, faster charging times, and upgrades in UTM, may be required to assist those potential marketplace developments [12].

7.2.6 Phase 6: Point-to-point air taxi services

The emergence of "air taxi" services, which provide on-demand near-point services with varying infrastructure based on city density and flight demand, marks the end of the most advanced phase. Infrastructure and capacity limitations, weather-related reliability, noise, safety, air traffic issues, and other impacts associated with distributed and large-scale operations make true point-to-point aviation in urban environments. Transport may be theoretically impossible. The next part examines the prospects and hurdles of implementing UAM to commercial viability [13].

7.3 UAM Architecture

7.3.1 UAM as a system of systems

System of systems (SoS), a term recently formalized by ISO 21839:2019 [9], is a collection of systems that interact and provide unique capabilities that cannot be achieved by individual subsystems alone. The approach to understanding complex systems in areas such as renewable energy, national security, infrastructure, transportation, and defense is called systems of systems (SoS) [14].

Figure 7.3 shows the components or systems of interest in urban air mobility. These include vehicles, heterogeneous fleets, vertiport systems, flight operations (trajectory, dispute resolution, and safety), passenger demand, energy systems, and their life cycles at the vehicle, fleet, and network-level elements. These systems are independent in terms of management and operation. The configuration of component systems creates an SoS that requires collaboration to function correctly and experience its excellence

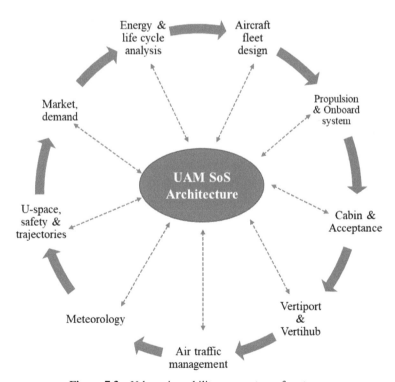

Figure 7.3 Urban air mobility as a system of systems.

efficiently. Moreover, each design is geographically dispersed and evolves independently over time (e.g., introducing new vehicles, new energies, smart grids, new ATM processes, weather or meteorological phenomena, etc.). UAM aircraft and related technologies are the systems of interest (SoI) covered in this chapter [15].

7.3.2 Infrastructure, vehicle, and network Modeling

Most PAV concepts require takeoff and landing infrastructure and are subject to the same regulations and specifications as traditional heliports. The idea of a PAV suitable for driving describes the ability to take off and land on common road infrastructure [16]. Location and capacity are highlighted as critical aspects of the VTOL infrastructure. As such, the UAM extension contains the following attributes of the VTOL landing site: UAM station. The different properties of the UAM stations and UAM vehicles are listed as follows:

- A specific identifier (e.g., a unique string or sequential number)
- Status (i.e., MAT Sim network connection to which the station is connected)
- Cargo capacity (i.e., capacity for simultaneous VTOL operations)
- A unique identifier (e.g., a unique string or sequential number)
- Original location (i.e., unique identifier for overnight parking at UAM station)
- Passenger capacity (i.e., the maximum number of simultaneous passengers)
- Cruising speed (i.e., maximum horizontal airspeed)
- VTOL speed (i.e., maximum vertical ascent/descent speed)
- Maximum range (i.e., maximum distance between launch station and landing station)
- Start and end of service (i.e., earliest and latest time of vehicle use)

The various properties of the nodes and links in an UAM network is listed as follows:

- A distinguishing mark (e.g., a unique string or sequential number)
- Place and altitude (i.e., x, y, and z coordinates)
- Special identifiers (such as unique strings or sequential numbers)
- Source and destination nodes (i.e., unique identifiers for connected nodes)
- Length (i.e., distance to fly between connected nodes)

- Throughput potential (i.e., the number of vehicles that can pass through the connection within a predefined time frame)
- Maximum free flow speed or maximum allowable or practicable cruising speed

7.3.3 Operation modeling

UAM vehicles are dynamically assigned to UAM stations during the simulation and routed accordingly. An agent must depart from a new location, register to the created departure handler with an activity or facility, and claim a UAM vehicle. When an agent request is placed, the dispatcher allows the UAM vehicle to be nearly available to the facility using the UAM router. Before the agent reaches the UAM station, the UAM vehicle may reach the departure UAM station upon placing the request. This activity is denoted as agent leg access. If no UAM vehicle is available, the request is queued, and the UAM dispatcher tracks if any vehicle becomes available before the parting time. The agent and the car wait for the other if either fails to arrive at the scheduled time [17].

The authorized departure handler uses DVRP's passenger engine to provide a way to book transportation for a specific location and time. While the current booking implementation (first available, closest UAM vehicle) is straightforward, bookings can be made at any point in the simulation, opening the door to the later development of more complex booking, allocation, and dispatch algorithms. Figure 7.4 illustrates how an agent is serviced using a UAM router for a ride. Access, UAM, and egress leg refer to the agent's navigation from/to the UAM station and access point.

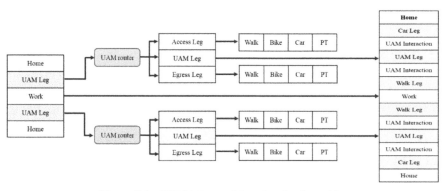

Figure 7.4   UAM inter-modal router plan for a ride.

Initially, all the UAM vehicles are registered to the DVRP package during the simulation implementation process. The UAM vehicle is characterized with the following tasks namely:

- UAM stay task (vehicle must maintain a current connection from the defined start and end times).
- The UAM fly task extends the DVRP drive task. This includes the flight path and a list of requirements that the vehicle must meet.
- UAM drop-off task (this includes passenger drop-off start and end times and location (via a network connection)).
- UAM pickup task, including start and end times and location (network connection) for picking up passengers.

Passenger pooling can also be added depending on the facility needed if more agents start and end at the same UAM stations. This facilitates the increase in passenger count and optimizes the passengers' cost and traveling time.

7.3.4 Framework of UAM

A modular and extensible SoS framework to assess UAM aircraft and fleet levels is depicted in Figure 7.5. The framework allows subsystem and system-level (aircraft) input to be passed to the SoS level considering operational scenarios and use cases. Aircraft-level inputs include architecture, cruise speed, cargo, reserve power requirements, and mission size. In addition, analysis of the lift-to-drag ratio and disc loading is possible. Factors such as batteries and charging technology are considered in the aircraft subsystem [18].

7.4 Simulation Models

7.4.1 Agent-based simulation

The internal arrangement of agent-based simulation (ABS) for analyzing a complex SoS is discussed in this section. For the UAM use case, the existing ABS framework is expanded. The demand model and the agent model are the two basic models that make up the ABS for UAM, with additional methods and classes provided for implementing the needed features (see Figure 7.6). The modularity of the simulation framework was prioritized during development; as a result, it effortlessly imitates any desired city, region, or country [19].

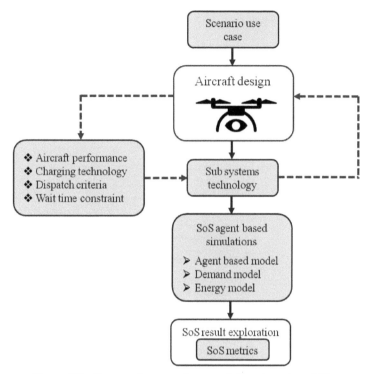

Figure 7.5 System of systems framework for urban air mobility.

When a demand at vertiport is raised by an agent based on the GPS coordinates, the request is generated. Specifying GPS and the generated number as input enables the automatic retrieval of the map of the desired region. From the details and availability of nearby UAM, the service of the UAM is decided. The infrastructure systems, specifically the vertiports, can be defined by users entering their locations or automatically locating places like subway stations or airports.

7.4.2 Demand model

The following explanation uses the words "demand" and "passenger" inter-changeably. A non-passenger carrying flight flown to reposition the aircraft is a deadhead flight or mission. Therefore, a "deadhead demand" is a deadhead assignment. A revenue flight or mission is also a flight that carries passengers. The demand model was developed specifically to ensure that it is modular to enable easy change of the demand model depending on the available data.

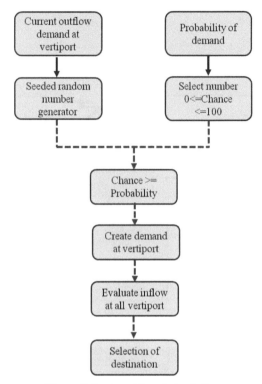

Figure 7.6   Agent-based simulation.

Inflow and outflow demand curves established at each vertiport comprise the demand model. The outflow demand curve governs the demand for departures from a vertiport [20]. On the other hand, the inflow-demand angle determines the need for incoming journeys to a vertiport. The inflow and outflow curves are defined as shown in Figure 7.7.

A demand curve is defined by a combination of several standard curves that form the desired demand distribution. The initial step is to get the value of the discredited outflow demand at the simulation time and convert it to a probability of demand generation at each time step. Then, a seeded random number generator chooses a chance value between 0 and 100 to compare this likelihood against. The vertiport generates an outbound demand if the chance value exceeds the possibility. The induced demand's final destination is determined using a weighted selection of the vertiport and considering the inflow demand magnitudes during the simulation time for each vertiport. This

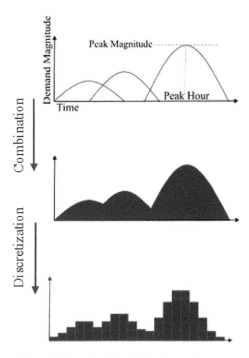

Figure 7.7 Definition of the demand curves.

means that a vertiport is more likely to be chosen as the destination if there is a higher demand at that vertiport during a specific period. From this point on, assigning and dispatching agents to complete the mission is the responsibility of the agent model. In conclusion, the inflow demand curves are used to determine the destination of the created demand, whereas the outflow demand curves are utilized to generate demand at a vertiport [21].

The aircraft awaits dispatch by the dispatcher once the assignment has been assigned. When the dispatch parameters are met, and the assigned mission of the aircraft agent is a revenue mission, the aircraft agent is dispatched. A successful dispatch occurs when the aircraft is fully loaded or when the wait time meets the desired target. The aircraft agent must wait to be assigned any potential new markets, raising the load factor. The aircraft agent is sent out immediately if the assignment is a deadhead mission in the unique scenario where the agent is given a deadhead mission as their next assignment. Provided that they are flying from and to the same vertiports, a demand can be issued to this deadhead mission before the dispatch of the deadhead flight. As a result, there may occasionally be passengers on a flight initially intended to be a deadhead flight [22].

7.4.3 Energy model

The purpose of the energy model in simulation is to track and update energy consumption and reload. Power consumption for turnover and deadhead missions is calculated in the sizing tool. Revenue mission power consumption is calculated for each possible load factor. In addition, the energy consumed in each flight state of the simplified mission profile, i.e., hover and cruise flight states, will be available in the simulation. Each simulation iteration updates the aircraft's available energy based on flight conditions and load factors. A simplified assumption of constant power consumption within each flight condition and load factor is employed. Aeroplane agents are also assumed to charge their batteries continuously, as defined by the charge power input in Birchport. Ninety percent charging efficiency is considered. All aircraft agents are supposed to start the simulation with a fully charged battery. More precisely, each agent has energy reduced from its maximum battery energy capacity by a percentage of its available power and the energy required to fly a primary mission [22].

7.5 Integration of Aircraft Design and Agent-based Simulation

First, a design of experiments (DoE) is created using each input parameter. Each DoE design point is sized for a specific aircraft architecture. The aircraft design tool then generates an extended DoE containing additional performance and aircraft parameters for each viable aircraft architecture, along with the inputs required to set up the simulation.

The ABS DoE runner module then runs DoE with multiprocessing and outputs the results. The DoE entries performed are summarized in Table 7.1.

7.5.1 Limitations of the UAM system of systems framework

This paradigm represents the first attempt to evaluate UAM use cases from the subsystem level to the system of interest to the system's design. Some restrictions that currently apply are:

- The framework assumes that each aircraft agent in the fleet is of the same type and has the same properties, such as cruise speed, payload, etc.

Table 7.1 Design inputs for experimentation.

Parameter	Count	Specific design points
Scenario	2	Near-term, far-term
Sizing mission	2	Single flight, multi-flight
Aircraft architecture	4	Multirotor, compound helicopter, lift + cruise, tiltrotor
Cruise speed, m/s	3	25, 40, 55
Passenger capacity	2	2, 4
Charging power, kW	3	250, 500, 1000
Passenger demand	2	Low (a max of 24 per hour), high (a max of 48 per hour),
Fleet size	9	12, 18, 24, 30, 36, 42, 48, 54, 60
Vertiport capacity	1	100 (unlimited)

Heterogeneous fleet simulations are technically possible, but dispatch and deadhead modeling still need to be improved.

- The framework allows unlimited birchport landing spots, gate/parking capacity, and notional passenger demand. A more thorough study of the market and barty port allows us to consider the demand of UAM passengers. Due to limited parking and landing areas, the schedule should be optimized further.
- The expected trajectory between vertiports is a direct flight with hovering and cruise times only. The trajectory model should be more complex and closer to what is expected in practice.
- The basic aircraft design process is constrained by assuming predetermined values for each aircraft architecture.
- Percentage of disk load, lift-to-force ratio, and curb weight from other predetermined properties and so on.
- The ratio of lift to drag remains constant and special attention should be paid to changes in cruising speed within the proposed framework.

This means a given aircraft architecture may only sometimes be perfectly replicated in its real-world counterpart rather than using established aircraft characteristics for sizing and modeling. Future research, therefore, focuses on more complex modeling of aerodynamics and weight. UAM encounters various obstacles in its development, including those related to safety and regulatory environment, air traffic control, noise, public acceptance, weather, environmental impact, infrastructure, and security. This section describes many potential roadblocks that can hinder the widespread adoption of UAM and some solutions that can be used to avoid them.

7.5.2 Case study for the validation of the SoS framework

The market or passenger demand is required to assess each city's UAM SoS. The authors chose to assume direction, which is determined parametrically based on the transport trends at the vertiport sites, as there needs to be a comprehensive study available and because of the inherent unpredictability of the market's acceptance of UAM. When the research is made public, this assumed data will be replaced with expected demand statistics for UAM operations. Additionally, vertiport locations should be determined based on multimodal transportation, demand, and city topography. For the sake of this study, the vertiport positions are presumptive, but latitude and longitude coordinates allow for simple modification [22].

The northern German city of Hamburg is an example in this case study. Hamburg, one of Germany's most congested cities, is a model for adopting UAM. The city's river may encourage early adoption when safety is considered. Assumed demand and vertiport locations explain the demand distribution at six vertiports. The vertiports are spaced apart on average by about 15 km.

7.5.3 MAS and NetLogo-based UAM

Some of the complications with present methodologies are the difficulty of testing scenarios on numerous cars with various variables in the test and determining the impact of particular aspects on the entire system. Additionally, simulations are the only natural way to collect data before creating a straightforward design without a genuine operating system. In this case, the strategy is to simulate several cutting-edge cars and build models with the essential characteristics to obtain accurate data. For the reasons mentioned, this strategy was the most effective way to deal with this problem.

A popular paradigm for creating distributed intelligent systems with or without agent cooperation is the multi-agent systems (MAS), a subset of distributed artificial intelligence. Approach challenging issues methodically. One of the essential paradigms in artificial intelligence is the agent-orientation paradigm, which allows many agents to operate freely in predefined environments. Numerous applications, including those that necessitate exceptional reliability, such as B. Aviation, are mentioned in the literature (Wooldridge, 2002). For MAS to be effective, several agents must interact and communicate with one another [23].

NetLogo is an agent-based language with a straightforward structure developed by Wilensky in 1999. You can create various programs of varying complexity using this language. The NetLogo with multi-agent capabilities keeps the logo's simplicity in versions with 2D and 3D simulation capabilities even though it adds more simulation capabilities. His NetLogo has been used to construct his MAS models in several academic works, including those in the social sciences and physics.

7.5.4 Model validation

Aviation experts considered each proposed scenario and compared the results to their predictions to validate the model. The following validation results were generated to support the study, as all landing and launch sites are probabilistically defined.

- Average vertical distance: The average vertical distance between the cruising altitude and the landing and takeoff locations for all simulated aircraft.
- Average horizontal distance: The average horizontal distance between landing and destination for all aircraft in the simulation.
- Average flight time: This is the average time that all aircraft spend at cruising altitude.
- Average vertical flight time: The average amount of time that all aircraft spend ascending.
- Average total time: The average times each aircraft uses.

Once the agent (airplane) behavior and generated outputs were determined to be consistent with the selected performance and other parameters, the model was considered validated and ready for simulation.

7.5.5 Tiers of complexity in agent-based modeling

The general overview of how agent-based systems have been developed for various dynamic urban processes, with a propensity to concentrate on specific components of urban systems where transportation plays a significant role. The agent-based models that are effective for modes of transportation are the emphasis of the section after this one, particularly those that use network capabilities to comprehend distances and methods of the vehicle thoroughly. In contrast to the complete model outlined in Section 7.2, the model addressed here concentrates on something other than a car. These models prioritize transportation over other jobs or services. This study is not focused on agent targeting or interactions [23].

Giving a gauge of the conceptual complexity included in transport models distinguishes models and makes comparing and contrasting them easier. Therefore, only the most elementary transport semantics are present in a model where objects or agents move freely inside a continuous space or grid. They convey space and have a fundamental understanding of distance. It is what it is: complexity level 0. Models classified as Tier 1 are those in which network behavior and other topological features are established to comprehend interactions better. Therefore, the agent would be more desirable if a house selection model could, in addition to knowing that a house is close to a public transportation stop, also consider the midpoint or centrality of that stop in the network. Knowing the result leads to the improvement of its properties. Tier 2 models are more demanding. In this model, network edges are weighted according to cost or benefit [23].

The cost may be reasonable. A model that minimizes the relative costs of toll roads fails to capture the impact of highways on the overall system adequately. Since agents may or may not have specific characteristics, prices can also be scheduled based on individual charges. Access to cars and public transportation expands or shrinks your network of agents. These models can have different strengths and weaknesses for each edge, including pollution-related costs and time and monetary costs. Tier 3 models at the top of the scale are the most advanced. It contains information about edges with weights in different simulation dimensions [23]. Agents can use the public transport network to reach their destinations slower but more comfortably by avoiding exceptionally crowded commuter trains. These Tier 3 models cover all behaviors that agents can imitate. It is provided here to point the direction of further research on agent-based modeling and transport models. For reference, the tier values are shown in Table 7.2.

7.6 Proposed Models

7.6.1 Heterogeneous multi-entity collaboration for UAM

Numerous market analyses, operational concepts (Con Ops), and other research studies describe the diversity of UAM's vision. Stakeholder enablement and proper management of UAS, including active support companies, cities, emergency service providers, regulators, and the general public as illustrated in Figure 7.8. As such, the FAA, NASA, and several business

Table 7.2 Tier levels and qualities.

Tier	Qualities	Examples
0	Surroundings with no restrictions on movement.	On a lattice, agents travel between homework and other location types steadily.
1	Topologically constrained environment, such as one that is networked or contains exclusion zones.	Along a network of roads, trains, and walkways, agents commute between their places of residence, employment, and other locations constantly.
2	Environmental topological structures with weights that limit mobility.	Agents travel between home, work, and other locations along a transportation network to minimize the economic cost function.
3	The topological structure's weights change depending on the simulation's various dimensions.	To minimize a cost function with economic and temporal components, agents move between their homes, places of employment, and other locations along a network with actuating track levels.

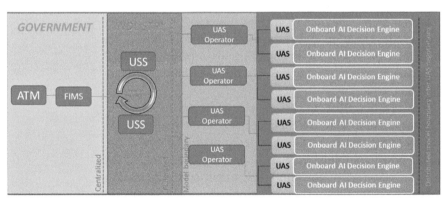

Figure 7.8 A distributed model for UTM utilizing edge computing capability on-board UASs.

partners are working together to design and build UTM systems. The UTM architecture presents a layered/hierarchical framework for managing urban air traffic, with organizations including UAS operators, supplemental data service providers, and UAS service suppliers collaborating with the UASs to make operational decisions. For UAM operations, it is envisioned that the various industry players – including UAS operators, USSs, and SDSPs – will coordinate with one another. The flight information management system is a vital component that guarantees the overall airspace safety used by conventional air vehicles like airplanes, helicopters, and UAS agents (FIMS). A central data exchange gateway for ATM, UTM, public service, and regulatory bodies is called FIMS. The use is authorized for functionality, quality of service, and other criteria connected to the FIMS in the UTM architecture.

FIMS UTM functions include registration and licensing, ensuring fair use of airspace, sharing information on aircrew restrictions and notices (NOTAMs), identification and authentication of data/service providers, and general airspace conditions and status. Surveillance and the possibility of conflict resolution of flight plans from different USS serving the same airspace, most importantly. However, UTM's effectiveness depends on the deployment paradigm for these tasks. For instance, considering tactical deconfliction in a centralized model may improve control and become a significant bottleneck as operation volume increases. On the other hand, the need for standardization can make a federated model complex.

Note that the UTM architecture incorporates the idea of vehicle-to-vehicle (V2V) communication capabilities, allowing the UAS to provide relevant information as an operator (e.g., four-dimensional proximity to the UAS). Recognize the situation and respond appropriately. NASA and others have studied the importance of maintaining air separation. Vehicle technology is designed to anticipate and resolve conflicts based on common intent in ADS-B and similar data.

The work described in this chapter is hoped to improve our ability to communicate information to enhance our collective situational awareness and reason by using that information to learn more about how the situation will develop. When combined with an appropriate onboard AI decision engine, such information will allow the UASs to put corrective measures into place either passively through proactive action or inter-UAS conversations. Certain circumstances call for direct UAS–UAS discussion with little USS participation for approval since USS–USS negotiation may be slow, unsafe, or useless in acquiring vehicle utilities. Because these USS are planned to be highly automated bots, latencies are anticipated to be significantly lower than they would be for human-centric operations. Nevertheless, most risks, such as communication network congestion, latency, brownout, etc., are reduced by the UAS decision-making level.

7.6.2 Data aggregation for AI-based cooperative control

A multi-agent, multi-nodal decision-making system must include data sharing, aggregation, and interpretation as essential elements. In the past, a wide range of research methodologies has been examined for decision-making in 4D trajectory management in ATC (three positioning dimensions and one-time dimension). Examples include trajectory-based air traffic control, mixed operations with airborne self-separation assurance, arrival sequencing, and the weather-cell avoidance solver known as auto resolver. Other examples include:

- Distributed trajectory flexibility preservation scenario complexity metric formulation.
- 4D trajectory planning.
- Human–machine interfaces for flight decks.

This study proposes a comprehensive data synthesis method into graphical data frames that convey information about the agents' behavior and intent along with the agents' biological data, enhancing the rich technological foundation. The opportunity to apply cutting-edge deep learning techniques for quick assessment and classification of conflicts in 4D traffic scenarios is the driving force behind such a data aggregation method.

Today, air traffic controllers use various visual tools to monitor and control airspace (and ground facilities like airports) traffic. These visual aids are intended to enhance human intellect with information that is simple to understand, allowing the human controller to make quick judgments to ensure efficiency and safety. The data input system must be tailored for the AI in the distributed UTM model, where the AI makes decisions. Data aggregation for autonomous entities like UASs must be AI-focused so that enormous collections of real-time and historical data may be efficiently synthesized and processed inside the vehicles.

7.7 Summary

Due to the high population density in metropolitan areas, research, business, and the government are struggling to solve urban transportation problems. This has created a new set of challenges for the UAM (urban air mobility) concept and its long-term viability. UAM is recommended for use in vehicles with vertical takeoff and landing (VTOL) capabilities and in electric variants known as eVTOL. The architectural design and model simulation are in progress. A complete usage for the public will be made available effectively in the future. Current AI technologies and data synthesis methods will be further explored through trait learning to establish the most efficient approaches for synthesizing the collective behavior of agents in different contexts. Implementation of the above technologies on wearable edge devices, such as NVidia Xavier Edge GPU modules with TensorRT optimization, is to be explored to enable scenario classification and integration of decision-making engines powered by UAM agents.

References

[1] Q. Shao, R. Li, M. Dong and C. Song, "An Adaptive Airspace Model for Quadcopters in Urban Air Mobility," in IEEE Transactions on Intelligent Transportation Systems, 2022, doi:10.1109/TITS.2022.3219815.

[2] N. Amilia, Z. Palinrungi, I. Vanany and M. Arief, "Designing an Optimized Electric Vehicle Charging Station Infrastructure for Urban Area: A Case study from Indonesia," 2022 IEEE 25th International Conference on Intelligent Transportation Systems (ITSC), 2022, pp. 2812-2817, doi:10.1109/ITSC55140.2022.9922278.

[3] A. S. Suzuki and Q. V. Dao, "A Flight Replanning Tool for Terminal Area Urban Air Mobility Operations," 2022 IEEE/AIAA 41st Digital Avionics Systems Conference (DASC), 2022, pp. 1-7, doi:10.1109/DA SC55683.2022.9925838.

[4] S. Bharadwaj, S. Carr, N. Neogi and U. Topcu, "Decentralized Control Synthesis for Air Traffic Management in Urban Air Mobility," in IEEE Transactions on Control of Network Systems, vol. 8, no. 2, pp. 598-608, June 2021, doi:10.1109/TCNS.2021.3059847.

[5] F. T. Souza and W. S. Rabelo, "A data mining approach to study the air pollution induced by urban phenomena and the association with respiratory diseases," 2015 11th International Conference on Natural Computation (ICNC), 2015, pp. 1045-1050, doi:10.1109/ICNC.201 5.7378136.

[6] R. Gillani, S. Jahan and I. Majid, "A Proposed Communication, Navigation & Surveillance System Architecture to Support Urban Air Traffic Management," 2021 IEEE/AIAA 40th Digital Avionics Systems Conference (DASC), 2021, pp. 1-7, doi:10.1109/DASC52595.2021.95943 79.

[7] J. -H. Woo et al., "AirScope: A Micro-Scale Urban Air Quality Management System," 2008 IEEE Fourth International Conference on eScience, 2008, pp. 378-379, doi:10.1109/eScience.2008.125.

[8] J. Bazurto, W. Zamora, J. Larrea, D. Muñoz and D. Alvia, "System for monitoring air quality in urban environments applyng low-cost solutions," 2020 15th Iberian Conference on Information Systems and Technologies (CISTI), 2020, pp. 1-6, doi:10.23919/CISTI49556.2020.9 141042.

[9] O. Ivashchuk and I. Ostroumov, "Separation Minimums for Urban Air Mobility," 2021 11th International Conference on Advanced Computer Information Technologies (ACIT), 2021, pp. 633-636, doi:10.1109/AC IT52158.2021.9548355.

[10] D. Helbing, I. Farkas and T. Vicsek, "Simulating dynamical features of escape panic," Nature, Letter, vol. 407, pp. 487-490, 2000.

[11] S. Boyd, "Convex optimization of graph Laplacian eigenvalues," Proceedings of the International Congress of Mathematicians, vol. 3, pp. 1311-1320, 2006.

[12] Y. LeCun, Y. Bengio and G. Hinton, "Deep Learning," Nature, vol. 521, no. 7553, pp. 436-444, 2015.

[13] L. Zhu, F. Guo, R. Krishnan and J. W. Polak, "A Deep Learning Approach for Traffic Incident Detection in Urban Networks," in 2018 21st International Conference on Intelligent Transportation Systems (ITSC), Maui, Hawaii, 2018.

[14] J. Bai and Y. Chen, "A Deep Neural Network Based on Classification of Traffic Volume for Short-Term Forecasting," Mathematical Problems in Engineering, 2019.

[15] L. Fridman, J. Terwilliger and B. Jenik, "DeepTraffic: Crowdsourced Hyperparameter Tuning of Deep Reinforcement Learning Systems for Multi-Agent Dense Traffic Navigation," in 32nd Conference on Neural Information Processing Systems (NIPS), Montréal, Canada, 2018.

[16] X. Ma, Z. Dai, Z. He, J. Ma, Y. Wang and Y. Wang, "Learning Traffic as Images: A Deep Convolutional Neural Network for Large-Scale Transportation Network Speed Prediction," Sensors, vol. 17, no. 4, pp. 818-833, 2017.

[17] K. He, X. Zhang, S. Ren and J. Sun, "Deep Residual Learning for Image Recognition," in Computer Vision and Pattern Recognition (CVPR), 2016.

[18] K. Simonyan and A. Zisserman, "Very Deep Convolutional Networks for Large-Scale Image Recognition," in 3rd International Conference on Learning Representations, San Diego, CA, 2015.

[19] C. Szegedy, V. Vanhoucke, S. Ioffe, J. Shlens and Z. Wojna, "Rethinking the Inception Architecture for Computer Vision," in Proceedings of the

IEEE conference on computer vision and pattern recognition, Las Vegas, NV, 2016.

[20] Mathworks, "Deep Learning with Images," 2018. [Online]. Available: https://www.mathworks.com/help/deeplearning/deep-learning-with-i mages.html.

[21] J. Snoek, H. Larochelle and R. P. Adams, "Practical Bayesian Optimization of Machine Learning Algorithms," Advances in neural information processing systems, pp. 2951-2959, 2012.

[22] NVidia, "Jetson AGX Xavier," 2019. [Online]. Available: https://www. nvidia.com/en-us/autonomous-machines/embedded-systems/jetson-a gx-xavier/.

[23] NVidia, "NVidia TensorRT," 2019. [Online]. Available: https://develope r.nvidia.com/tensorrt.

8

Reinforcement Learning Approaches for Urban Air Mobility/Navigation and Traffic Control Systems

**Sango Woo Jeon[1], J. Dharani[2], Min Dugki[1],
and Vishnu Kumar Kaliappan[2]**

[1]Department of Computer Science and Engineering, Konkuk University,
South Korea
[2]Department of Computer Science and Engineering, KPR Institute of
Engineering and Technology, India
E-mail: ndrw5580@gmail.com; dharanijagan92@gmail.com;
dkmin@konkuk.ac.kr; vishnudms@gmail.com

Abstract

Urban air mobility (UAM) is one of the emerging state-of-the-art approaches
for decision support systems in transportation. In the context of UAM, the
reinforcement learning (RL) approach can be used to develop navigation
and traffic control policies that maximize safety, efficiency, and passenger
satisfaction. One of the challenges that arise in developing RL-based UAM
navigation and traffic control systems is the complexity of the environment.
UAM involves multiple agents, such as drones, helicopters, and flying cars,
which need to navigate through a dynamic and uncertain environment. RL-
based algorithms are able to handle this complexity and uncertainty to
develop effective policies. Another challenge is the need for safety and
reliability in UAM systems. RL algorithms must learn safe and reliable
policies to handle unexpected events and disturbances. In this chapter, we
have investigated different RL approaches and environments used to control
the traffic systems in the UAM model.

Keywords: Urban air mobility (UAM), reinforcement learning (RL), policies, navigation and traffic control, multi-agent system

8.1 Introduction

The rise of UAM services such as drones, flying cars, and helicopters has increased the demand for effective navigation and traffic control systems. These systems must ensure the safety and dependability of UAM operations while increasing efficiency and passenger satisfaction. However, the UAM environment's complexity and dynamic nature present a significant challenge for developing effective navigation and traffic control systems [1]. RL has emerged as a promising solution to this problem. RL is a type of machine learning in which an agent learns to make decisions by interacting with its surroundings. RL algorithms have been applied successfully in various fields, including robotics, gaming, and finance [1]. RL algorithms can be used in the context of UAM to develop navigation and traffic control policies that optimize safety, efficiency, and passenger satisfaction.

The primary advantage of RL is its ability to deal with complex and dynamic environments. Learning from experience and adapting to changes in the environment allow RL algorithms to develop effective policies in uncertain and dynamic situations. Furthermore, RL can handle multi-agent interactions, critical for UAM systems involving multiple agents, such as drones, flying cars, and helicopters. Despite these benefits, developing effective RL-based UAM navigation and traffic control systems is difficult. Some key challenges that must be addressed are the complexity of the UAM environment, the need for safety and reliability, and the presence of multiple agents.

This chapter introduces RL approaches for UAM navigation and traffic control systems. The challenges of developing RL-based UAM systems and recent research in this field are discussed. Finally, the future directions for RL-based UAM navigation and traffic control systems will also be discussed.

8.2 Deep-Reinforcement-based Approach

With the advent of reinforcement learning, the controlling technology of UAM met a new paradigm. Cases and studies applying reinforcement learning are actively underway. In particular, it is possible to intelligently control multiple aircraft by applying multi-agent deep reinforcement learning. In this chapter, we will learn about the basic concepts of reinforcement learning, various reinforcement learning algorithms, and multi-agent reinforcement learning algorithms. In addition, we will look at actual cases and related studies applying these technologies to navigation and traffic control systems for UAM.

8.2.1 Deep reinforcement learning

Deep reinforcement learning (DRL) is a field that has recently been spotlighted in the machine learning field. This technology involves the acquisition of a model by an agent through a process of trial and error within a specified environment, without the utilization of any pre-existing data. Reinforcement learning can be described as a learning process that develops behavior through trial and error to maximize the cumulative reward in a sequential decision-making problem, and this sequential decision-making problem can be expressed by the Markov decision process (MDP). Reinforcement learning can be used in various fields such as stock investment, driving, and games.

8.2.1.1 Markov decision process (MDP) and Markov game

Reinforcement learning is an optimization method for solving sequential decision-making problems using the MDP [2]. By adding the concept of reward, Markov reward process (MRP) is defined as follows.

$$(S, A, P, R, \gamma)$$

where S stands for *state* space, and A stands for *action* space. P is the probability distribution of the following state s' when the agent chooses the action $a \epsilon A$ from the state $s \epsilon S$, and R means the reward received in the next state s. For the cumulative reward, the future reward is depreciated using the

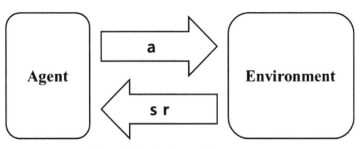

Figure 8.1　Markov decision process.

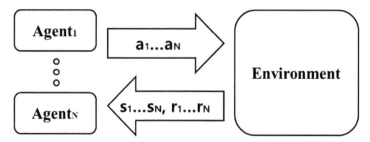

Figure 8.2   Markov decision process environment and multiple agents.

discount rate γ. This discount rate reflects future uncertainty and prevents the divergence of the cumulative reward so that learning can be performed stably.

Markov game is a multi-agent extension of MDP. The Markov Decision Process is illustrated in the Figure 8.1. Markov game is defined as a set of states and actions for N agents, a probability distribution for the next state given through each agent's current state and action, and a reward function for each agent that depends on the global state and action of all agents. Observation Oi is a partial state that agent i can observe and includes some global state information. Each agent learns the policy $\pi : Oi \rightarrow (Ai)$ that maximizes the expected sum of rewards. The multiple Agent MDP process is shown in the Figure 8.2.

8.2.1.2 Components of reinforcement learning

Reinforcement learning agents consist of a model for the environment, policy function, and value function. The model of the environment is the agent's prediction of what the next state of the environment and reward will be. It consists of two components, which are the state model and the reward model. The state model is the probability distribution of the next state when the current state and action are given.

$$\mathcal{P}_{ss'}^{a} = P\left[S_{t+1} = s' | S_t = s, A_t = a\right]. \tag{8.1}$$

The reward model is the reward expectation when current state and action are given

$$\mathcal{R}_s^a = \mathrm{E}\left[R_{t+1} | S_t = s, A_t = a\right]. \tag{8.2}$$

The policy is an agent's behavior pattern. It tells what action to take in a given state. In other words, it is a function that associates the state with the action. The policy is divided into deterministic policy and stochastic policy. The deterministic policy gives an action for a given state, and stochastic policy gives a probability distribution of actions for a given state.

$$a = \pi(s) \tag{8.3}$$
$$\pi(a|s) = P\left[A_t = a | S_t = s\right] \tag{8.4}$$

The value function is a function that predicts how much reward the state and action will return later when following the policy. That is, it is the weighted sum of all rewards to be received after taking the corresponding state and action. At this time, the discounting factor is used to indicate preference for the reward to be received before the reward to be received later.

Policy evaluation:

$$v_\pi(s) = E_\pi[R_{t+1} + \gamma R_{t+2} + \gamma^2 R_{t+3} + \ldots | S_t = s] \tag{8.5}$$
$$v_\pi(s) = E_\pi[G_t | S_t = s]. \tag{8.6}$$

When an action is given, the value function is called an action value function and is expressed as q(s,a). It takes action and state as inputs.

$$q_\pi(s, a) = E_\pi[G_t | S_t = s, A_t = a]. \tag{8.7}$$

8.2.1.3 Bellman equation

Solving MDP can be said to solve prediction and control problems. Prediction is a problem of evaluating the value of each state, given a policy. Control is the problem of finding the optimal policy. To solve these problems, in other words, to find the optimal value and policy, the policy and the value can be expressed through the Bellman equation. The value function and the action value function can be expressed as the Bellman expectation equation and the Bellman optimal equation [3]. Bellman's equation uses the recursive relationship between the present time t and the next $t + 1$.

8.2.1.3.1 Bellman expectation equation

Bellman expectation equation represents the expectation value of v(s) and q(s,a) when following given policy .

$$v_\pi(s_t) = E_\pi[R_{t+1} + \gamma v_\pi(s_{t+1})] \tag{8.8}$$
$$q_\pi(s_t, a_t) = E_\pi[R_{t+1} + \gamma q_\pi(s_{t+1} a_{t+1})] \tag{8.9}$$
$$v_\pi(s) = \sum_{a \in A} \pi(a|s) q_\pi(s, a) \tag{8.10}$$
$$q_\pi(s, a) = R_s^a + \gamma \sum_{s' \in S} P_{ss'}^a v_\pi(s') \tag{8.11}$$

$$v_\pi(s) = \sum_{a \in A} \pi(a|s) \left(R_s^a + \gamma \sum_{s' \in S} P_{ss'}^a v_\pi(s') \right) \tag{8.12}$$

$$q_\pi(s, a) = R_s^a + \gamma \sum_{s' \in S} P_{ss'}^a \sum_{a' \in A} \pi(a'|s') q_\pi(s',a'). \tag{8.13}$$

8.2.1.3.2 Bellman optimal equation

Bellman's equation is an equation for optimal values. The optimal value can be defined as follows:

$$v_*(s) = \max_\pi v_\pi(s) \tag{8.14}$$

$$q_*(s, a) = \max_\pi q_\pi(s, a). \tag{8.15}$$

The optimal value is calculated when following optimal policy, which makes maximum cumulative rewards.

$$v_*(s) = v_{\pi_*}(s) \tag{8.16}$$

$$q_*(s, a) = q_{\pi_*}(s, a) \tag{8.17}$$

$$v_*(s_t) = \max_a E\left[R_{t+1} + \gamma v_\pi(s_{t+1})\right] \tag{8.18}$$

$$q_*(s_t, a_t) = E[R_{t+1} + \gamma \max_{a'} q_*(s_{t+1}, a')] \tag{8.19}$$

$$v_*(s) = \max_a q_*(s, a) \tag{8.20}$$

$$q_*(s, a) = R_s^a + \gamma \sum_{s' \in S} P_{ss'}^a v_*(s') \tag{8.21}$$

$$v_*(s) = \max_a \left[R_s^a + \gamma \sum_{s' \in S} P_{ss'}^a v_*(s') \right] \tag{8.22}$$

$$q_*(s, a) = R_s^a + \gamma \sum_{s' \in S} P_{ss'}^a \max_a q_*(s',a') \tag{8.23}$$

8.2.1.4 Reinforcement learning algorithm

The first distinction that distinguishes reinforcement learning algorithms is the existence of a model for the environment. The second distinction between reinforcement learning algorithms is whether to use value functions and policies. A value-based agent has the advantage of using data more efficiently.

A well-known example of a value-based algorithm is deep Q-network (DQN). On the other hand, policy-based agents have the advantage of learning more stably because they optimize directly for what they want. The reinforcement algorithm is a representative example of a policy-based algorithm. There are not only two extreme cases. Some agents have both a value function and a policy, called an actor−critic agent.

8.3 Deep Q-Network

DQN [4] was introduced in *Playing Atari with Deep Reinforcement Learning* on NIPS in 2013 and human-level control through deep reinforcement learning on Nature in 2015. It is evaluated as the first case of successful application of deep learning to reinforcement learning. The frame of the game is used as input data, and the action of the agent comes out as an output. In the experiment, trained agent surpassed human in Atari 2600 games (Figure 8.3).

DQN is value-based learning, which does not use explicit policy. In other words, DQN only uses a value function trained by a neural network and selects an action. The policy is implicit such as selecting the action with the highest value. The algorithm is based on Q-learning algorithm. Bellman optimal equation is used to train optimal action-value q^* in Q-learning. Q

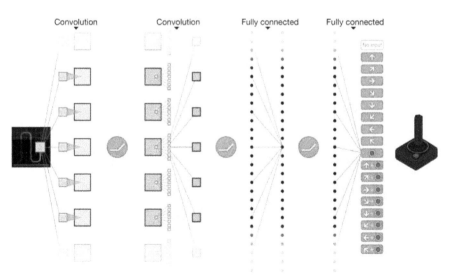

Figure 8.3   Atari games.

value is updated by reducing the value between the estimated q value from the current state and the estimated q value from the next state. The estimated q value from the next state is considered a target since it is closer to the optimal value from eqn (8.19).

$$q_* (s_t, a_t) = E_{s'}[R_{t+1} + \gamma \max_{a'} q_*(s_{t+1}, a')]$$
(8.24)

where $q(s, a) \leftarrow q(s, a) + \alpha(R + \gamma \max_{a'} q(s', a') - q(s, a))$.

The q function using a neural network parameter θ can be expressed as follows:

$$q_\theta (s, a).$$
(8.25)

A loss function of the network is defined as the square of the difference between the target and $q_a(s, a)$.

$$L(\theta) = E\left[(R + \gamma \max_{a'} q_\theta (s', a') - q_\theta(s, a))^2\right].$$
(8.26)

The parameters are updated in the direction of reducing the loss through gradient descent.

$$\theta' = \theta + \alpha(R + \gamma \max_{a'} q_\theta \left(s', a'\right) - q_\theta (s, a) \nabla_\theta q_\theta (s, a).$$
(8.27)

The DQN algorithm overcomes two problems in deep reinforcement learning.

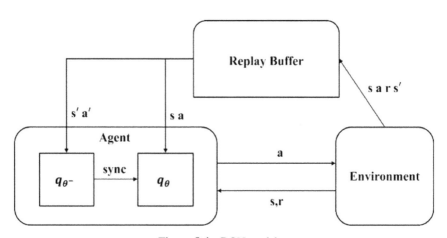

Figure 8.4   DQN model.

One is the problem of highly correlated data. In order to train a neural network, data samples should be independent. In a reinforcement learning environment, data are related because the data from the current and next states are highly related. A replay buffer was introduced to solve this problem. As shown in the Figure 8.4. The agent collects its experience called transition in the replay buffer. The transition consists of a set (s, a, r, s'). In the training phase, randomly selected data from the replay buffer is used. In this way, the correlation is reduced.

The second problem is non-stationary data distribution. Whenever the network is updated, the agent's behavior will be changed. Then distribution of the data changes, which hinders training. To solve this problem separated target network θ^- is used.

$$L(\theta) = E\left[\left(R + \gamma \max_{a'} q_{\theta^-}\left(s', a'\right) - q_{\theta}(s, a)\right)^2\right]. \qquad (8.28)$$

At regular intervals, the target network parameter θ^- is synchronized with the θ. This makes the target stationary and stabilizes the training.

8.3.1 Reinforcement

Reinforce is a basic policy-based reinforcement learning algorithm. It uses policy gradient theorem to reinforce its policy. Before updating policy network $\pi(s, a)$, we need to define the objective function $J(\theta)$ to evaluate the policy. The $J(\theta)$ is defined as the expected value of the sum of rewards

$$J(\theta) = E_{\pi_\theta}\left[\sum_t r_t\right] \qquad (8.29)$$

$$= v_{\pi_\theta}(s_0)$$

$$= \sum_{s \in S} d(s) * v_{\pi_\theta}(s). \qquad (8.30)$$

To improve the policy network, we need to maximize the objective function. To achieve this goal, gradient ascent method is needed.

$$\theta' \leftarrow \theta + \alpha * \nabla_\theta J(\theta). \qquad (8.31)$$

To calculate $\nabla_\theta(\theta)$, several steps are required as policy gradient theorem.

From eqn (8.29), .

$$J(\theta) = \sum_{s \in S} d(s) * v_{\pi_\theta}(s)$$

$$= \sum_{s \in S} d(s) \sum_{a \in A} \pi_\theta(s, a) * R_{s,a} \tag{8.32}$$

$$\nabla_\theta J(\theta) = \nabla_\theta \sum_{s \in S} d(s) \sum_{a \in S} \pi_\theta(s, a) * R_{s,a} \tag{8.33}$$

$$= \sum_{s \in S} d(s) \sum_{a \in S} \nabla_\theta \pi_\theta(s, a) * R_{s,a} \tag{8.34}$$

$$= \sum_{s \in S} d(s) \sum_{a \in S} \frac{\nabla_\theta \pi_\theta(s, a)}{\pi_\theta(s, a)} * \pi_\theta(s, a) * R_{s,a}. \tag{8.35}$$

Since

$$\frac{d(\ln x)}{dx} = \frac{1}{x}, \quad \frac{\nabla_\theta \pi_\theta(s, a)}{\pi_\theta(s, a)} = \nabla_\theta \log \pi_\theta(s, a)$$

$$E_{\pi_\theta}[\nabla_\theta \log \pi_\theta(s, a) * R_{s,a}] \tag{8.36}$$

$$\approx G_t * \nabla_\theta \log \pi_\theta(s, a) \tag{8.37}$$

Reinforce algorithm

Initialize policy function $\pi_\theta(s, a)$ with random weights
for episode 1 to M do
   initialize state s to s_0
   following π_θ, generate an episode $s_0, a_0, r_0, s_1, a_1, \ldots, s_{T-1}, s_T, a_T, r_T$
   for each step of the episode t = 0, ..., T

$$G_t \leftarrow \sum_{i=t}^{T} r_i * \gamma^{i-t}$$

$$\theta \leftarrow \theta + \alpha * G_t * \nabla_\theta \log \pi_\theta(s, a)$$

8.3.2 Actor−critic

The actor−critic algorithm is a methodology that uses a policy function and a value function together. The actor selects an action based on the policy, and the critic evaluates the policy based on the value function. In the agent's learning process, the policy and value networks are learned, respectively. The critic network is updated to learn the value of the current policy function.

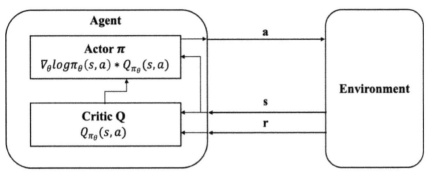

Figure 8.5   Actor–critic model.

Figure 8.5 illustrates the flow of Actor–Critic model. The actor-network is trained in such a way that, through evaluation of the value function, if the result is good, it is reinforced and if it is not good, it is weakened.

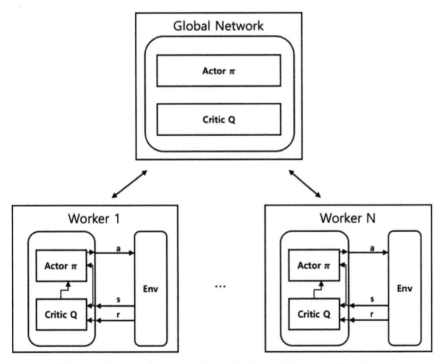

Figure 8.6   Asynchronous methods for deep reinforcement learning.

Some modifications are made to improve the performance of the actor−critic algorithm. This is the Q-actor−critic algorithm.

$$\nabla_\theta J\left(\theta\right)=E_{\pi_\theta}[\nabla_\theta\log\pi_\theta\left(s,a\right)*Q_{\pi_\theta}(s,a). \tag{8.38}$$

In the advantage actor−critic algorithm, an advantage function $A\pi\theta(s, a)$ is used to take a specific action compared to the average, general action at the given state

$$A_{\pi_\theta}\left(s,a\right)=Q_{\pi_\theta}\left(s,a\right)-V_{\pi_\theta}(s,a). \tag{8.39}$$

One of the popular actor−critic algorithms was introduced in *Asynchronous Methods for Deep Reinforcement Learning 2016*. In this paper, asynchronous method of training advantage actor−critic algorithm was introduced. It runs multiple agents instead of one and updates the shared network periodically and asynchronously as shown in Figure 8.6.

8.4 Multi-agent Reinforcement Learning

8.4.1 Multi-agent reinforcement learning algorithms

Single-agent reinforcement learning [5], a general reinforcement learning algorithm, is expressed in Figure 8.7. In reinforcement learning, with the given, agent, and the environment, the learning is accomplished through the repetition of the agent's environmental state observation, action performed by the agent, the change of the environment according to the agent's action, and the reward payment by the environment.

As time passed, researchers tried to solve more complex situations through reinforcement learning, and they are actively researching the field of multi-agent deep reinforcement learning (MADRL) [1], a technology in

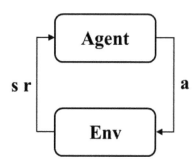

Figure 8.7   Single agent reinforcement learning.

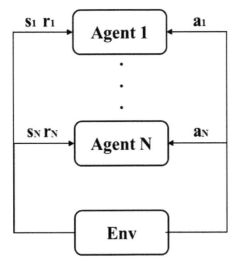

Figure 8.8 Multi-agent reinforcement learning algorithms.

which multiple agents perform specific goals, such as in a soccer game. In MADRL, as shown in Figure 8.8, each agent receives different observations and rewards from the same environment and takes different actions. One challenging part of the field of MADRL is that the environment is not stationary from the point of view of one agent because multiple agents act simultaneously and affect the environment. Overcoming these problems is essential for improving the performance of the MADRL algorithm.

Multi-agent reinforcement learning is typically derived from the techniques of single-agent reinforcement learning to suit multi-agent environments. However, there are many ways for modification, such as having an independent policy function and value function, use of centralized learning way of communication between agents, use of centralized training and decentralized execution, or applying various methods of communication between agents. This section will examine several multi-agent learning algorithms to understand its concept and methods.

8.4.2 Multi-agent deep deterministic policy gradient

Multi-agent deep deterministic policy gradient is a multi-agent deep deterministic policy gradient (MADDPG), which combines deterministic policy gradient (DPG) with DQN approaches such as replay buffer and target separation. Each agent has its own actor and critic. Centralized critic approach

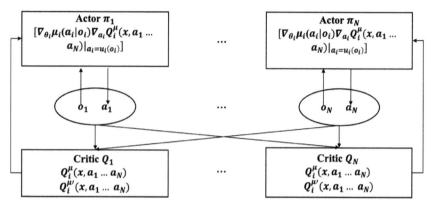

Figure 8.9 Centralized critic approach.

is used when training each critic, which means that observations and actions from all other agents are used as depicted in Figure 8.9.

8.4.3 Multi-actor-attention-critic

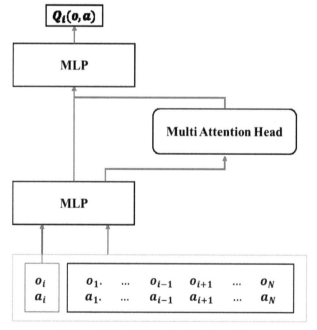

Figure 8.10 Multi-actor-attention-critic.

Multi-actor-attention-critic is a MADRL algorithm that trains decentral-ized policies in multi-agent settings, using centrally computed critics that share an attention mechanism that selects relevant information for each agent at every timestep. Each agent has its own policy and critic and takes action within the environment through independent judgment. The centralized critic approach helps each agent to learn critic by using all agent's actions and observations to overcome the challenge of a non-stationary environment. The Attention function in the multi-attention head can be described as mapping a query and a set of key-value pairs to an output, in which the output is computed as a weighted sum of the values, where a compatibility function of the query with the corresponding key computes the weight assigned to each value. Through the multi-attention head layer, each agent receives a value-weighted to important information among other agents' information. This attention mechanism enables more effective and scalable learning in complex multi-agent environments compared to recent approaches. Figure 8.10 clearly illustrates the architecture of Multi Actor attention critic.

In multi-actor-attention-critic (MAAC) algorithm, an attention mecha-nism that selects relevant information for each agent at every timestep was added.

8.5 Application for UAM

This section explores the application of reinforcement learning and multi-agent reinforcement learning in navigation and traffic control systems, as demonstrated by previous research studies.

8.5.1 Recent work

Research related to navigation and traffic control systems for UAM can be found on logistic delivery or passenger transportation problems. This involves the problem of intelligently controlling multiple aircraft at the same time as each aircraft autonomously flies. In early works, single-agent reinforcement learning-based algorithms were applied. In the 2020s, you can find many studies based on multi-agent reinforcement learning that can be applied to an environment where multiple drones cooperate to perform a specific goal.

The reinforcement-learning-based drone delivery model is developed by the experiment shown in Figure 8.11, in which double DQN (DDQN) is employed to guide a single drone to deliver the cargo. DDQN is an algorithm that solves the action value overestimation problem of DQN. The approach is simulated using the Airsim simulation engine retrieves and uses virtual

Title of the paper	Algorithm used	Remarks
(A) Reinforcement learning for drone delivery	DDQN	Applied in assigning passenger to drone or vehicle
(B) Reinforcement learning to optimize the logistics distribution routes of unmanned aerial vehicle	Policy Gradient	Achieve path planning methods for the UAV (unmanned aerial vehicle) in goods delivery
(C) Joint optimization of multi- UAV target assignment and path planning based on multi-agent reinforcement learning	MADDPG	Simultaneous target assignment and path planning (STAPP)
(D) MADRL-based drone taxi control	QMIX	eVTOL vehicles that can compute the optimal passenger transportation routes
(E) Multi-agent reinforcement learning-based UAS control for logistics environment	MAAC	Applied in cooperative delivery task

Figure 8.11 Experiment setup using Airsim [7].

sensor data such as LiDAR, stereo vision camera including depth information, and IMU. In this study, a network called joint neural network (JNN), a combination of a dense layer of the state scaler with the existing convolutional neural network (CNN) based network, is adopted.

In Figure 8.12, the policy gradient method for the path planning of UAV distribution is proposed. This research focused on finding the optimal route for delivering multiple customers. The model found the optimal strategy of visiting 50 nodes in 200 iterations, which is very efficient. However, there is a risk of collision with other drones since the situation in which flying with other drones together is not considered.

The multi-agent deep reinforcement learning approach was used in the framework shown in Figure 8.13 to address the collaborative control system for UAVs. A novel framework utilizing the multi-agent deep deterministic policy gradient (MADDPG) algorithm was developed to address the optimization challenge of target assignment and path planning for a large number of unmanned aerial vehicles (UAVs). The problem is formally defined in the

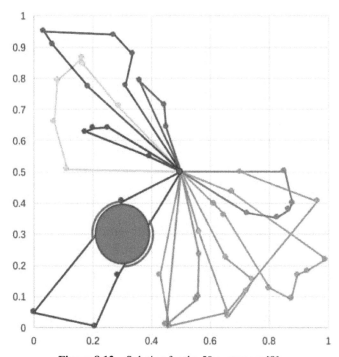

Figure 8.12 Solution for the 50 customers [8].

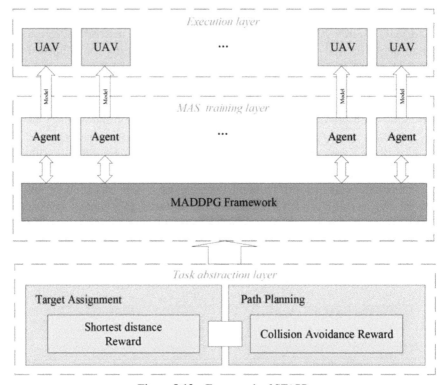

Figure 8.13 Framework of STAPP.

following manner. UAVs effectively circumvent the designated threat area and successfully reach their target by strategically minimizing travel distance. The quantity of target and unmanned aerial vehicles (UAVs) is equivalent. The problem at hand can be subdivided into two distinct subproblems, namely target assignment and path planning. The rewards aim to minimize the distance traveled and the number of collisions.

Application for the drone taxi using electric vertical takeoff and landing vehicle (eVTOL) was introduced in 2021. From Figure 8.14, the scenario is that multiple eVTOLs with limited battery capacity transport passengers while charging batteries from a specified charging station. Q-Mix-based multi-agent reinforcement learning is used to manage multiple vehicles. Q-Mix is multi-agent reinforcement learning in which q learning and centralized learning and decentralized execution are applied.

Improved MAAC (I-MAAC) [6] is applied for logistic delivery as shown in Figure 8.15. In this work, cooperative logistic delivery service virtual

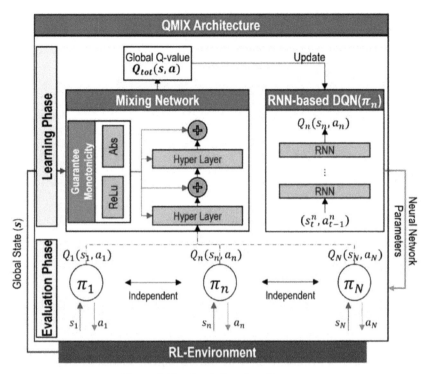

Figure 8.14   Q-mix based architecture.

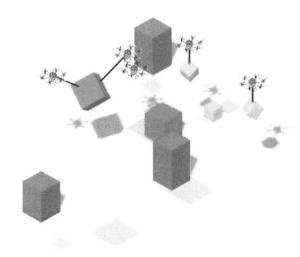

Figure 8.15   UAS LDS environment.

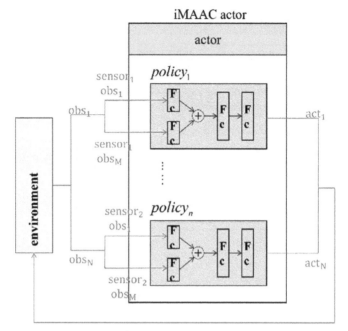

Figure 8.16   Actor of the I-MAAC.

environment (UAS-LDS) using Unity is introduced. Collaborative tasks are explicitly shown in this study. In the scenario, multiple UASs should deliver two types of cargos. One cargo type needs one UAS, and the other cargo type should be carried by two UAS.

Virtual LiDAR sensor called Ray cast sensor supported by the Unity, GPU, and IMU is used. The actor network is modified to reflect multiple types of sensor and the result was improved than the original MAAC in Figure 8.16.

8.5.2 Challenges

Recent studies have shown that there are still clear limitations in solving UAM application problems through reinforcement learning [11]. Most of the research is applying and verifying the algorithm through a virtual environment, which means that there are parts that are difficult to apply in the real environment. Also, even in the virtual environment, sophisticated flight dynamics still need to be reflected, and the virtualized sensor is different from the real sensor. The first difficulty in applying it to the real environment is the communication between agents. Delays and overheads that occur when multiple agents communicate with each other while flying and send and receive information in real time can be a problem. Another difficulty is predicting what kind of problem will appear when an algorithm trained in a virtual environment is applied to an actual situation. In order to overcome these problems, SILS (software-in-the-loop) and HILS (hardware-in-the-loop) technology should be developed. Also, a detailed digital twin environment should be needed.

8.6 Summary

Efficient navigation and traffic control systems ensure safety, dependability, and passenger satisfaction in urban air mobility (UAM) services. Reinforcement learning (RL) has emerged as a promising approach to developing effective UAM navigation and traffic control systems. In order to formulate efficacious policies within contexts characterized by uncertainty and dynamism, it is imperative to engage in a comprehensive and systematic approach. RL algorithms can handle the complexity of the UAM environment and adapt to changes in the environment. Developing effective RL-based UAM navigation and traffic control systems, on the other hand, presents significant challenges, including the need for safety and reliability, as well as the presence of multiple agents. Despite these obstacles, recent research has yielded promising results, such as developing drone collision avoidance policies, optimizing helicopter routes, and enhancing passenger pickup and drop-off processes. More research is needed to address the complexity, safety, and reliability challenges in UAM systems.

References

[1] Kaelbling LP, Littman ML, Moore AW. Reinforcement learning: A survey. Journal of artificial intelligence research. 1996 May 1;4:237-85.
[2] Bellman R. A Markovian decision process. Journal of mathematics and mechanics. 1957 Jan 1:679- 84.
[3] O'Donoghue B, Osband I, Munos R, Mnih V. The uncertainty bellman equation and exploration. InInternational Conference on Machine Learning 2018 Jul 3 (pp. 3836-3845).
[4] Roderick M, MacGlashan J, Tellex S. Implementing the deep q-network. arXiv preprint arXiv:1711.07478. 2017 Nov 20.
[5] Papoudakis G, Christianos F, Schäfer L, Albrecht SV. Benchmarking multi-agent deep reinforcement learning algorithms in cooperative tasks. arXiv preprint arXiv:2006.07869. 2020 Jun 14.

[6] Jeon, Sangwoo, Hoeun Lee, Vishnu Kumar Kaliappan, Tuan Anh Nguyen, Hyungeun Jo, Hyeonseo Cho, and Dugki Min. 2022. "Multiagent Reinforcement Learning Based on Fusion-Multiactor-Attention-Critic for Multiple-Unmanned-Aerial-Vehicle Navigation Control" Energies 15, no. 19: 7426.

[7] Muñoz, G., Barrado, C., Çetin, E., & Salami, E. (2019, September 10). Deep Reinforcement Learning for Drone Delivery. Drones, 3(3), 72. https://doi.org/10.3390/drones3030072.

[8] Feng, L. (2020). Reinforcement Learning to Optimize the Logistics Distribution Routes of Unmanned Aerial Vehicle. ArXiv. /abs/2004.09864

[9] Liu, Z., Qiu, C., & Zhang, Z. (2022). Sequence-to-Sequence Multi-Agent Reinforcement Learning for Multi-UAV Task Planning in 3D Dynamic Environment. Applied Sciences, 12(23), 12181 https://doi.org/10.3390/app122312181.

[10] Jo, H., Lee, H., Jeon, S., Kaliappan, V. K., Anh Nguyen, T., Min, D., & Lee, J. W. (2022, September 30). Multi-agent Reinforcement Learning-Based UAS Control for Logistics Environments. Lecture Notes in Electrical Engineering, 963–972 https://doi.org/10.1007/978-981-19-2635-8_71.

[11] Salagame, A., Govindraj, S., & Omkar, S. N. (2022). Practical Challenges in Landing a UAV on a Dynamic Target. ArXiv. /abs/2209.14465

9

Challenges in Charging of Batteries for Urban Air Mobility

K. Mohana Sundaram[1] and B. Kavya Santhoshi[2]

[1]KPR Institute of Engineering and Technology, India
[2]Godavari Institute of Engineering and Technology (A), India
E-mail: kumohanasundaram@gmail.com; kavyabe2010@gmail.com

Abstract

While researchers have been making great progress in developing batteries to power everything from passenger cars to segments of the electric grid, employing batteries to power aircraft capable of carrying people has remained a major challenge. In this chapter, the various challenges faced using batteries for fast charging and durable operation is discussed.

Keywords: Urban air mobility, unmanned vehicle, fast charging, battery

9.1 Introduction

Battery energy density is one of the most critical design parameters for sizing all-electric aircraft, but it is easily overestimated. Establishing the effective, usable energy density is confused by varying degrees of margin needed to account for structural and thermal management between different cell chemistry and pack designs. A better methodology is needed to fairly compare emerging battery technologies for electric aircraft. Currently, there is a loss of critical information when vehicle trade studies are performed using "nominal" published cell-level performance metrics. Aircraft power demands rarely match these nominal power profiles, and aircraft designers lack the ability to accurately simulate the battery performance and temperature off-nominally unless the battery chemistry is well established [1]. Conversely, battery suppliers are unable to publish more realistic performance metrics due to a lack of generalized reference cases [2]. This can lead to poor assumptions. For example, aircraft studies may assume a fixed discharge efficiency of a battery, when in reality the usable energy in a pack is dependent on the power and thermal profile. Information needed to properly assess weight penalties for thermal management is also typically poorly characterized when assessing candidate batteries.

9.2 Battery Thermal Management System

The use of EV will increase in near future and so priority is given to the need of developing effective batteries. The thermal degradation of the batteries is a big challenge for better BTMS, which affect the range of the EV. The main objective of the BTMS is to control the temperature of the battery cell and thus improve the battery life. Li-ion batteries are usually preferred for their energy storage in electric vehicle. There are many challenges such as low efficiency at high and low temperatures, decreased life of electrodes at high temperature and the direct effect on the performance, reliability, cost, and protection of the vehicle, and the safety issues related to thermal runaway in lithium-ion batteries. So an effective thermal battery management system is therefore one of the most crucial technology for long-term success of an electric vehicle. Normally the temperature ranges from 25 to 40 °C are the optimal working conditions for the Li-ion batteries. When the temperature of these batteries is higher than 50 °C, it degrades the life of the battery [3].

9.3 Batteries for Electric Vehicle in Indian Market

Barriers for EVs in the Indian market can be addressed from various prospectives such as technical barriers, policy barriers, and lack of infrastructure.

9.3.1 Vehicle servicing

In order to take proper care of the electric car, a trained technician should be available to repair, maintain, and find troubleshoot of the electric vehicle. They must be able to apply their skills to rectify the problem as quickly as possible.

9.3.2 High capital cost

The battery packs of an electric vehicle are expensive, and also it needs replacement more than once in its lifetime. The gas-powered cars are cheap when compared with electric vehicles. Consumer perception plays a vital role in attracting new customer and retains an existing customer. Despite the growing range in the auto market with a broader range of electric vehicles, the choice of buying an electric car is limited and is expected to continue over time. So there should be awareness of the company offerings to the customer by means of advertising, social media, or another channel. Studies show that the lack of knowledge associated with the government scheme, economic benefit, and awareness of the vehicular technology can have a direct impact on the electric vehicle adoption.

9.3.3 Raw materials for batteries

The raw materials for EVs batteries include lithium, nickel, phosphate and manganese, graphite, and cobalt, which are rare earth materials. For an internal combustion engine, aluminum copper and steel are required. The catalyzers for combustion automobiles need platinum, rhodium, and palladium to filter the toxic gases. These all are scare materials, and the availability of these materials may not be enough for battery production. The lithium-ion batteries alone consume 5 million tons/year of nickel, which could lead to 10–20 times more consumption of lithium and cobalt in future.

9.3.4 Battery lifespan/efficiency

The electric cars are usually created by using electric motors, batteries, chargers, and controllers by replacing fuel tank and gasoline engine of a conventional vehicle. As the EV batteries are designed for a long life, it wears out in due course of time. Currently, most manufacturers are offering eight years/100,000 mile warranty for their batteries.

9.3.5 Driving range of electric vehicle

A driving range is recognized as the main barrier of an electric vehicle typically because EVs have a smaller range as compared with the equivalent ICE vehicle. The distance an electric vehicle can travel on a full charge or full tank is considered as a significant drawback to uptake the EV in the global market. Most of the BEV provides a driving range of less than 250 km per recharge [4]. However, some of the latest models can offer up to 400 km. By now, PHEV is offering a range of 500 km or more due to the availability of liquid fuel internal combustion engines. The driver must plan their trip carefully and may not have the option for a long-distance trip. This makes the magnitude of driving range as a barrier.

9.3.6 Charging time

Charging time is closely related to the issue of driving range. With a slow charger, the EV can take up to 8 h for a full charge from the empty state using a 7 kW charging point. The charging time mainly depends on the size of the battery. Bigger the size of car batteries, longer the time it takes to recharge the battery from empty to full state. Also, the charging time of the battery directly depends on the charging rate of the charge point. Higher the charging price of the charge point, lower will be the time taken by the battery to get fully charged. In the current scenario, rapid chargers are used to charge the vehicle in a faster way reducing the time required. The commercially available electric cars are compatibles with charge points having a higher maximum charge rate than they can handle. This indicates that the battery can be charged at a maximum rate that they can handle without any fault. However, the charging rate of the battery with rapid charger reduces with a decrease in temperature or at cold temperature. The EV chargers are categorized in accordance with their charging speed at which their battery gets recharged. There are three fundamental kinds of EV charging, for example Level 1, Level 2, and DC fast. Level 1 charging utilizes a standard 120 V outlet by converting AC to DC using an on-board converter. It takes 8 h to charge the EV with 120 V outlets for a range of approximately 120–130 km. Level 1 charging is basically done at home or in the working environment. Level 2 chargers are typically set up at a public place or workplace that can be charged with a 240 V outlet. It takes 4 h to charge the battery for a range of 120–130 km. With DC fast charging, the change from AC to DC occurs in the charging station that has the most fast charging arrangements. This permits stations to supply more power, charging vehicles in a quicker way. It can charge the battery in 30 min for a range of 145 km.

9.4 Safety Requirement

The electric vehicle must meet the safety standard as specified by state or local regulation. The batteries should also meet the testing standards that are subject to conditions like overcharge, temperature, short circuit, fire collision, vibration, humidity, and water immersion. The design of these vehicles should be such that they should have safety features like detecting a collision, short circuit, and should be insulated from high voltage lines.

9.5 Battery and Electric Grid Considerations

There is a fundamental tradeoff between battery size and mission length. On one hand, bigger batteries have more energy, which can translate into longer missions. However, with the increase in battery size comes additional aircraft weight, which can translate to increased acquisition cost. Based on the current battery technology, eVTOL aircraft will likely need to be partially or fully recharged after each mission. Given current battery-charging technologies, the time to perform this charging is likely to deter high aircraft utilization, particularly during peak demand periods. The amount of electricity required to power an electric fleet of aircraft is not trivial and will likely have significant impacts on the electric grid, which may not be able to be supported by the current electric grid. The issues are described by Kohlman and Patterson (2018) as follows: "If UAM vehicles are to be all-electric, as many are proposing, there will be new demands placed on the electrical grid infrastructure that must be understood. Additionally, vehicle-level characteristics such as the recharge time or energy used for a flight will have direct impacts on the efficiency, cost, and ultimate viability of UAM networks. For example, if vehicles must be charged for long periods of times between missions, a very large number of charging stations will be required at vertiports and many vehicles may be required to meet demand for UAM services."

Further, the cost of grid upgrades to support UAM operations is not trivial. A recent report by Black & Veatch estimates the cost to extend an existing service line to support 31 MW chargers to be between $75K and $100K; the cost for a new feeder line to support up to 83 MW chargers to be between $2.6K and $1.3M per mile; a new transformer bank over 10 MW to support over 15 chargers to be between $3M and $11M; and a new substation bank over 20 MW to support 30 chargers to be between $40M and $80M (Stith, 2020). The impact of charging on operations and the number of required charging stations has been noted by other authors. In a study of cargo operations in the San Francisco Bay Area, German et al. (2018) found that for a lift + cruise eVTOL concept model and a tiltrotor aircraft model, charging times with a 300 kW charger ranged from 12.5 to 19.1 min and 16.0 to 23.1 min, respectively. When the charger was increased to 400 kW, these charge times decreased to 9.5 to 14.4 min and 12.1 to 17.4 min, respectively. The impacts of UAM operations on the electric grid were clearly demonstrated in a study by Justin et al. (2017). Based on an examination of electric aircraft for regional distances, they generated power profiles for stations where Cape Air and Mokulele Airlines operate. Cape Air's network included 525 daily flights to 43 airports primarily in the New England area using mostly twin-engine piston-powered Cessna 402s. Mokulele's network included 120 daily flights to airports primarily in the Hawaiian Islands using 11 single-engine turboprop Cessna 208s. They found very high peak powers at the airlines' busiest airports, i.e., for Cape Air the peak power exceeded 1 MW in Nantucket Memorial (ACK) airport and in Boston Logan International (BOS) airport, which is the order of magnitude of the demand of approximately 1000 households. For Mokulele, the peak-power at Molokai airport (MKK) was 517 kW, which is about 1/20th of the total generation capability for the entire island of Molokai (Justin et al., 2017).

The authors explored various operational strategies to reduce peak-power demands and the cost of electricity, and found that a strategy that includes optimizing battery recharging with battery swaps can achieve reductions on the order of 20% compared to a power-as-needed strategy. What is clear from these and other publications is that the power requirements on the electric grid are not trivial, and significant opportunities exist to optimize the deployment of charging and fast-charging stations. Furthermore, given that electricity prices vary across cities and providers, the optimal battery recharging solution will likely be city-dependent.

9.6 Battery Recycling

The batteries used in electric vehicles are generally planned to last for a limited lifetime of the vehicle but will wear out eventually. The pricing for battery replacement is not properly informed by the manufacturers, but if there is a need for battery replacement outside its warranty period, then it adds the expenses by dumping the old battery with a new one. The chemical elements of the batteries like lithium, nickel, cobalt, manganese, and titanium not only increase the cost-effectiveness of the supply chain but also have environment concerns during scraping of the battery elements.

9.7 Charge Scheme Definitions

9.7.1 Slow charging

Slow charging is typically associated with overnight charging. This is a definition easy to grasp that translates into a 6- to 8-h period. Slow charging makes use of the EV or PHEV on-board charger, which is sized based on input voltage from the grid. For example, a 120 V, 15 A (80%) service would supply a 1.4 kW charger, while a 240 V, 32 A service would supply a 6.6 kW charger.

How does this translate into recharging the vehicle battery pack? A PHEV with a 5 kWh battery pack, for example, would have a 1.4 kW on-board charger that allows complete recharge on the order of 5 h. An EV with a 40 kWh battery pack might have a 6.6 kW charger, which allows complete recharging on the order of 6−8 h, depending on thermal considerations and charge algorithms for the battery chemistry.

9.7.2 Fast charging

Fast charging could be defined as any scheme other than slow charging. But the real definition, or set of definitions, is much more complex. Table 1 lists a few of the more commonly used terms, which include fast charge, rapid charge, and quick charge. The California Air Resources Board (ARB), in their Zero Emissions Vehicle (ZEV) mandate program, lists a certification requirement for fast charging as a 10-min charge that enables the vehicle to travel 100 miles.

9.8 Cycle Life, Charge Schemes, and Cost for Lithium Battery Chemistries

Batteries are the primary reason that EVs are not the vehicles we drive today. Cost and range issues have hampered mass adoption. The latest advances in lithium chemistries promise answers to these issues. Will it be enough to beat other transportation alternatives? Will it happen soon?

9.8.1 Iron phosphate, lithium titanate, but what is next?

Since the days of lead acid and nickel metal hydride batteries used in the GM EV-1, the search for batteries with higher energy density has been unending. In fact, hybrid vehicles such as the Toyota Prius and Honda Civic still use nickel metal hydride batteries. Nickel metal hydride, however, in addition to relatively low energy density, also contains nickel, an expensive material. Three of the more popular lithium-based electrochemical systems vying for EV applications are lithium cobalt or lithium manganese oxides (standard format), lithium iron phosphate, and lithium titanate.

Batteries such as those for laptop computers and cell phones, which use standard format lithium cells, would seem to be the obvious solution because they have a relatively low cost and very high specific energy. In fact, a concept developed by AC Propulsion using "bricks" of 18,650 computer cells configured into large packs has been adopted by both BMW (Mini Cooper) and Tesla Motors. Computer cells, which are manufactured by the billions in Asia, have disadvantages. Due to safety and performance reasons they are limited to 1- to 2-h charge rates at the pack level so as not to impact cycle life.

Lithium iron phosphate batteries are manufactured by many companies worldwide and have gained credibility through their use in power tools. Lithium iron phosphate cells have a much lower energy density than standard format cells, but can be charged much faster—on the order of 20−30 min.

Lithium titanate batteries allow charging on the order of 10 min and have been shown to have extremely long cycle life—on the order of 5000 full depth of discharge cycles. Lithium titanate has high inherent safety because the graphite anode of standard format and iron phosphate batteries is replaced with a titanium oxide.

9.8.2 Range vs. pack size

Batteries, even lead acid batteries, are expensive. The larger the battery pack, the more expensive it is. What is the right size for an EV battery pack? One measure of EV efficiency is Wh/mi on a plug-to-wheels basis. The lower the Wh/mi, the more efficient the EV drive train. A smaller, less expensive battery pack may one day allow the full functionality of today's conventional vehicles if fast charging is part of the equation.

9.9 Summary

Electric vertical takeoff and landing (eVTOL) aircraft have attracted considerable interest as a disruptive technology to transform future transportation systems. Their unique operating profiles and requirements present grand challenges to batteries. Research and interest in UAM have grown exponentially over the past five years, but significant questions remain with respect to whether UAM will become the next disruptive technology in urban transportation. As seen in the meta-analysis of UAM publications, much of the emphasis to date has been focused on fundamental questions [5]. How do we design an eVTOL aircraft? How can we create more energy-dense batteries to support eVTOL missions? How do we design the airspace so that high-volume eVTOL operations can occur simultaneously with commercial and drone operations? Will there be demand for an eVTOL air taxi service and, if so, which business cases make the most sense—commuting, business shuttles to an airport, or other trip purposes? In contrast, research in EV/AV and sharing technologies for ground transportation is further along, and researchers and communities have experiences in designing and implementing EV fleets, some of which are part of ride-hailing or car-sharing applications.

References

[1] K. Chaudhari, A. Ukil, K. N. Kumar, U. Manandhar and S. K. Kollimalla, "Hybrid Optimization for Economic Deployment of ESS in PV-Integrated EV Charging Stations," in *IEEE Transactions on Industrial Informatics*, vol. 14, no. 1, pp. 106-116, Jan. 2018.

[2] N. Liu, Q. Chen, X. Lu, J. Liu and J. Zhang, "A Charging Strategy for PV-Based Battery Switch Stations Considering Service Availability and Self-Consumption of PV Energy," in *IEEE Transactions on Industrial Electronics*, vol. 62, no. 8, pp. 4878-4889, Aug. 2015

[3] Straubinger, Anna, et al. "An overview of current research and developments in urban air mobility–Setting the scene for UAM introduction." *Journal of Air Transport Management* 87 (2020): 101852.

[4] Thipphavong, David P., et al. "Urban air mobility airspace integration concepts and considerations." *2018 Aviation Technology, Integration, and Operations Conference*. 2018.

[5] Bauranov, Aleksandar, and JasenkaRakas. "Designing airspace for urban air mobility: A review of concepts and approaches." *Progress in Aerospace Sciences* 125 (2021): 100726.

10

Safety and Security Challenges in Implementing Urban Air Mobility

Dhivya Rathinasamy[1], R. Sivaramakrishnan[2], R. Pavithra[3], and Vishnu Kumar Kaliappan[2]

[1]Department of Artificial Intelligence and Data Science, PSNA College of Engineering and Technology, India
[2]Department of Computer Science and Engineering, KPR Institute of Engineering and Technology, India
[3]Department of Computer Science and Engineering, NGP Institute of Engineering and Technology, India
E-mail: dhivya.rathinasamy@gmail.com;
sivaraamakrishnan2010@gmail.com;
pavithra.sri7@gmail.com; vishnudms@gmail.com

Abstract

Urban air mobility (UAM) is a new concept of transportation that has the potential to revolutionize urban mobility by providing efficient, fast, and environmentally friendly solutions for urban transportation. This chapter provides a comprehensive overview of UAM, its potential benefits, and challenges. It explores the need for UAM and the challenges it presents in terms of safety and security. The chapter discusses the different types of security measures, including passive, preventive, active, and destructive measures, and emphasizes the importance of comprehensive measures to address these challenges. Additionally, this chapter highlights the limitations of existing research on UAM and the need for further studies to fully understand the implications of UAM implementation in urban areas. While UAM has the potential to revolutionize urban transportation, it also presents significant challenges that must be addressed through continued research, planning, and collaboration between stakeholders to ensure its safe and sustainable implementation in the future.

Keywords: Cyber-attacks, detection methods, unmanned aerial vehicle

10.1 Introduction to UAM

The urban population has been rapidly increasing in recent years, primarily due to the economic upturn, rising job opportunities, and modernization, with more than 56.2% of the global population now urbanized [1]. Therefore, smart transportation systems play a crucial role in meeting the growing transportation needs. Urban air transportation is an on-demand transportation service for people that offers fast, manageable, and cost-effective transportation, making it the future transportation model. To meet the expected high demand, autonomy is necessary for a financially viable transportation system. As the amount of vehicle automation and autonomous vehicle operations increase, urban air mobility will be impacted. Intelligent systems are required for effective control and decision-making, which go beyond the capabilities of highly trained human operators. However, these systems are also vulnerable to cyber-attacks, which could have significant consequences. Therefore, an enhanced and secured control system is necessary for urban air mobility.

10.1.1 Evolution of UAM

The growth of urban populations has been notable in recent years due to economic growth, employment opportunities, and modernization. Early aviation services date back to 1909 when DELAG started the first airline service. Table 10.1 describes the various airline services from 1909 to 1971. Urban air mobility (UAM) services started in 1940, and the Los Angeles Airways operated a regional helicopter service from 1947 to 1971. Flying car concepts were also introduced during this period. However, during World War II, the

Table 10.1   Airline services.

Year	Service	Reference
1909	DELAG German Airship Transportation Corporation Ltd	https://www.airships.net/delag-passenger-zeppelins/
1916	Aircraft Transport and Travel Limited	https://www.britannica.com/topic/Air-Transport-and-Travel-Ltd
1953	New York Airways	https://www.nytimes.com/1979/05/16/archives/new-york-airways-acts-to-file-for-bankruptcy-suing-sikorsky.html
1971	Los Angeles Airways	https://www.latimes.com/visuals/photography/la-me-fw-archives-airways-helicopter-overturn-20170221-story.html

Axis forces used aircraft to crash into buildings, resulting in tragic civilian casualties.

Number of scheduled passengers boarded by the global airline industry from 2004 to 2022 (in millions) [2].

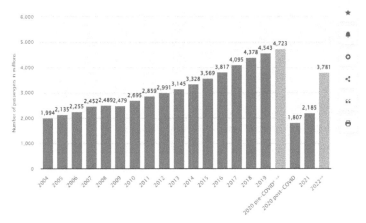

Figure 10.1.1   Increasing number of Passengers.

According to BIS Research, the analysis of the UAM industry indicates a substantial growth of 25.71% CAGR during the forecast period of 2023−2035. The Asia-Pacific region is predicted to be the dominant market by 2035 with a share of 40.10%. This region comprises China, Japan, South Korea, and Singapore, but it is projected that China will hold a significant share in 2035 due to the escalating population and traffic congestion in its megacities [3]. Figures 10.1.1 and 10.1.2 depict the increasing passenger usage and current market scenario based on mobility.

Global Urban Air Mobility Market Forecast, 2023-2035

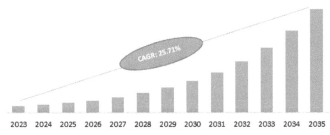

Figure 10.1.2   Global Urban air market forecast.

10.1.2 Main technologies

There are two main technologies used in UAM as described below.

1) **Autonomy:** In the context of UAM, the conventional approach to data processing is not recommended due to the need for real-time access to a large pipeline of data. UAM's capabilities can be expanded by increasing autonomy and incorporating a large volume of sensory data, which leads to reduced costs and increased safety. Drones and unmanned aircraft system (UAS) are examples of autonomous flights.

2) **Electric aircraft propulsion:** UAM will begin with the introduction of electric aircraft, which poses several challenges. The rapid evolution of digitization and electrification has a significant impact on UAM. The key components of this environment include [4]:

   a. **Autonomous self-piloted aircraft:**
   Autonomous self-piloted aircraft are aircraft that are capable of transporting passengers without human intervention, using automatic intelligent systems for operation and control.

   b. **Vertiport**
   Vertiports are airports designed for vertical takeoff and landing of aircraft. They come in two types: public use and private use. Education and training are provided to engineers and architects to enable them to provide these services.

   c. **Operation centers**
   Operation centers are the base for controlling and monitoring all data and processes related to various components of the aviation transportation system.

   d. **Maintenance centers**
   Maintenance centers are used to identify defects and take corrective actions, and they maintain a complete database for this purpose.

   e. **A governance structure**
   Governance structures are used to manage operations, policies, and objectives. They include finance management, accountability, business and risk assessment, safety assessment, and roles and responsibilities.

   f. **Consumer software applications**

The software facilitates various services such as automation, maintenance, and management. It is also utilized for inventory management and mobile deployment.

10.1.3 The underlying concepts

This section captures the focus on emerging technologies such as UAM, RAM, UAVs, and VTOLs, as well as the importance of innovative mobility solutions for the future. It covers their definitions, characteristics, and applications they present [4].

a) **Rural air mobility (RAM):**
   The concept is still in development and aims to create a safe, efficient, accessible, affordable, and versatile air transportation system for passenger and cargo mobility in rural and exurban areas.

b) **Urban air mobility (UAM):**
   This is a developing concept that envisions a secure, effective, accessible, economical, and versatile air transportation system for the mobility of passengers and the delivery of goods within urban areas or while traveling through them.

c) **Small unmanned aircraft:**
   This type of aircraft is commonly referred to as drones and has a maximum takeoff weight of less than 55 pounds, which includes all equipment and cargo on board or attached.

d) **Small unmanned aircraft system (sUAS):**
   This refers to the small unmanned aircraft and all of its necessary components including communication links and control systems that are needed to ensure the safe and efficient operation of the aircraft within the national airspace system.

e) **Unmanned aerial vehicles (UAV):**
   An unmanned aerial vehicle (UAV), also known as a drone, is a completely automated aircraft that does not require a human operator on board. Primarily used for military operations, UAVs can be classified based on weight, degree of autonomy, altitude, and other criteria. UAV architecture involves numerous sensors, actuators, and software to enable autonomous flight.

f) **Vertical takeoff and land (VTOL):**

This refers to a type of aircraft that has the ability to hover, take off, and land vertically, without the need for a runway or other takeoff and landing infrastructure.

10.1.4 UAM maturity scale

NASA's UAM Coordination and Assessment Team has created a UAM maturity level (UML) scale, which is a framework with various applications. Based

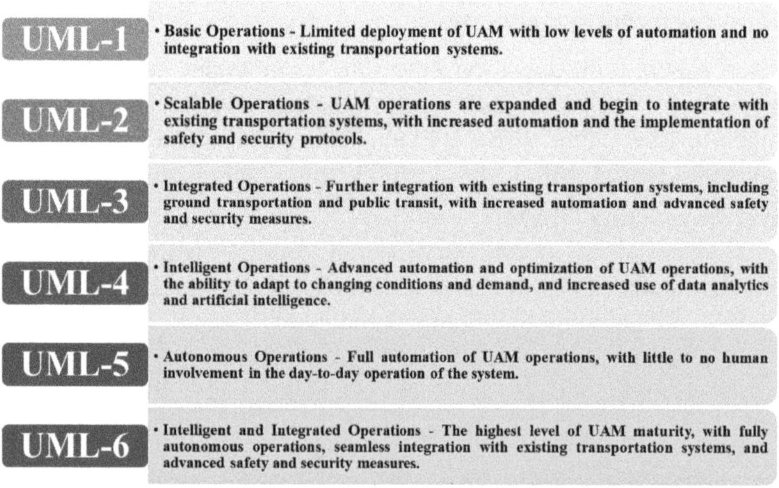

UML-1	• Basic Operations - Limited deployment of UAM with low levels of automation and no integration with existing transportation systems.
UML-2	• Scalable Operations - UAM operations are expanded and begin to integrate with existing transportation systems, with increased automation and the implementation of safety and security protocols.
UML-3	• Integrated Operations - Further integration with existing transportation systems, including ground transportation and public transit, with increased automation and advanced safety and security measures.
UML-4	• Intelligent Operations - Advanced automation and optimization of UAM operations, with the ability to adapt to changing conditions and demand, and increased use of data analytics and artificial intelligence.
UML-5	• Autonomous Operations - Full automation of UAM operations, with little to no human involvement in the day-to-day operation of the system.
UML-6	• Intelligent and Integrated Operations - The highest level of UAM maturity, with fully autonomous operations, seamless integration with existing transportation systems, and advanced safety and security measures.

Figure 10.2 UAM maturity level (UML) scale.

on Figure 10.2, which is representing the UAM maturity level (UML) scale, firstly, it provides an understanding of the expected operational capabilities of a UAM air transportation system as it develops over time. Secondly, it enables analysis of technology and regulatory requirements that are associated with the UAM maturation process. Thirdly, it allows for an evaluation of the current maturity of various segments of the UAM ecosystem. Fourthly, it facilitates the coordination of priorities and areas of emphasis within the UAM ecosystem. Lastly, it helps increase community and public awareness of UAM and its potential impact on mobility in the future.

10.1.5 Legal and societal stakes for UAM

UAM has the potential to revolutionize transportation systems by providing efficient, safe, and sustainable aerial transportation within urban areas. However, to implement UAM, a well-defined legal framework is essential to manage airspace operations above cities. The legal framework should consider the unique nature of the airspace, especially the very-low airspace located above cities. Therefore, non-aviation stakeholders such as city planners, policymakers, and urban designers must be included in the development of the legal framework.

The framework must be designed to ensure that UAM systems are integrated safely and efficiently with existing transportation systems, infrastructure, and land use. Additionally, the legal framework should ensure that

UAM operations meet societal goals such as reducing congestion, minimizing environmental impacts, and improving accessibility. Failure to develop a clear legal framework for UAM could result in delays, conflicts, and safety concerns, which could impede the progress of UAM systems. Therefore, it is essential to develop a legal framework that considers all stakeholders, including non-aviation stakeholders, to ensure the successful implementation of UAM systems [5].

10.1.6 Applications of UAM

UAM has numerous applications in various fields such as transportation, emergency response, search and rescue, agriculture, and construction. This section will focus on the most important applications of UAM, highlighting how it can bring about significant improvements in efficiency, safety, and sustainability.

a. **Construction:** Unmanned aerial vehicles (UAVs) are no longer limited to passive data collection tasks but are now being equipped with manipulators and robotic arms to perform active tasks that involve interacting with the environment [6]. These capabilities have extended the range of applications for UAVs, including contact-based inspection, manipulation, and grabbing and assembly. In addition, UAVs are currently being used for various purposes in the field of urban air mobility (UAM), such as collecting data before, during, and after the construction phase. This includes tasks such as building surveys and maintenance, health and safety inspections, progress tracking and reporting, security, building information modeling, and monitoring environmental factors.

b. **Agriculture:** Aerial services in agriculture have been found to offer numerous advantages, leading to their extension into other tasks such as top dressing, which involves the application of fertilizers over farmland from the air since 1940. In modern times, unmanned aerial vehicles are used for precision agriculture. Relevant applications of UAM in agriculture include evaluating nutrients and assessing health, analyzing water stress, estimating yield and biomass, monitoring soil, detecting weeds, monitoring the environment, and spraying crops from the air.

c. **Energy:** Maintenance and inspection of offshore wind turbines (OWTs) is a difficult task that involves a high degree of risk for manual operators working in hazardous environments. The complexity of the tasks involved has hindered the adoption of unmanned solutions [7]. However, drone-based inspection presents several advantages over traditional methods like rope-access inspection and ground-based camera inspection. The use of drones can significantly reduce the number of personnel required to travel to and climb up wind installations using crew transfer vessels or helicopters, which in turn reduces emissions. Drones can also eliminate the need for heavy lifting equipment during inspections, making the process more efficient [8].

d. **Entertainment:** The entertainment industry has a history of leading technological advancements. Recently, they have focused on incorporating drones or unmanned aerial vehicles (UAVs) into their operations. The use of drones has revolutionized aerial photography and videography, making it accessible even to amateur photographers. This new technology allows for stunning shots that were previously impossible to capture.

e. **Law enforcement:** According to a survey conducted by policeone.com in 2018, over 200 police officers in the United States were asked about the use of unmanned aerial vehicles (UAVs) in their departments. The top findings were as follows:

- 83% of respondents reported that UAVs were used for search and rescue operations.
- 79% reported that UAVs were used for disaster management purposes.
- 76% reported that UAVs were used for SWAT operations.
- 71% reported that UAVs were used for monitoring crime and traffic.

While drones have been adopted by law enforcement agencies worldwide, they also face various regulatory, ethical, and legal issues. Some of these issues include privacy concerns, data protection, safety regulations, and cybersecurity threats. Law enforcement agencies are also exploring the use of artificial intelligence and machine learning to enhance drone capabilities, such as autonomous flight and object recognition. Overall, the use of drones in law enforcement has both benefits and challenges that need to be carefully considered and managed.

f. **Other applications:** Drones have significant potential in the public health sector and this potential is continuously expanding. Some of the potential uses of drones in this field include the distribution of vaccines, transportation of blood, medicines and biologicals, assistance in medical emergencies and disaster relief efforts, transportation of organs for transplantation, and surveillance in difficult areas.

10.2 UAM Cyber-Attacks

Drones are now widely used for tasks like monitoring, search and rescue, and inspecting infrastructure. However, there are concerns about security and privacy because there are no clear rules for operating drones. Criminals can easily use drones because they are portable, cheap, and easy to fly [9]. Cyber-attacks are also possible which is represented in Figure 10.3, which can affect communication and the drone's onboard systems [10].

10.2.1 Channel jamming attack

RF jamming techniques can be used to disrupt the radio communication between the unmanned aerial vehicle (UAV) and ground control by increasing the level of noise interference at the RF receiver. There are various types of RF jamming techniques that can be employed [DW-4].

A summary of jamming techniques is shown in Table 10.2 [12].

Noise jamming is also referred to as "brute force jamming." It works by transmitting noise with the appropriate shape, power, and bandwidth to overlap with the jammed signal, preventing transmission between the transmitter and receiver. This increases the level of interference on the receiver side and decreases the signal-to-noise ratio (SNR) [13].

10.2.2 Message integrity and spoofing attack

a. **Message interception attack:** The unauthorized acquisition of confidential information can occur when an adversary intercepts sensor data. This type of attack can be either passive or active. In a passive attack, the malicious actor eavesdrops on communication links to obtain information [14].

b. **Message deletion/message injection:** The data or messages transmitted by UAVs are at risk of being deleted or manipulated by adversaries, which can lead to compromised data. This includes inspection data such as images and telemetry, as well as log files, which can be falsified [15].

c. **GPS spoofing attack:** A GPS spoofing attack involves the use of counterfeit GPS signals to replace genuine GPS signals and redirect a UAV toward a fake destination. This type of attack can also enable attackers to capture and manipulate UAVs through GPS spoofing, which could guide the UAV away from its intended route without raising any alarms. As a result of GPS spoofing, the true status of a UAV can be altered by attackers mentioned in Figure 10.4 [16].

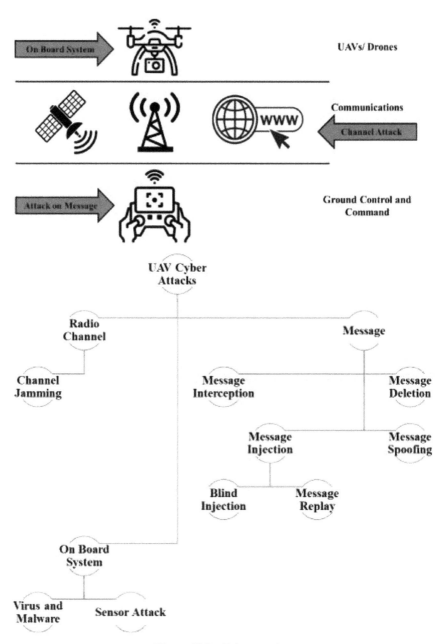

Figure 10.3   Cyber-attacks.

Table 10.2 Summary of jamming techniques.

Technique	Pros	Cons
Noise	The simplest form of jamming involves applying interference to a small portion of the radio frequency spectrum with the required power.	The jamming technique that applies a high level of power to a small portion of the spectrum is not effective, and it can be easily mitigated since it does not perform any dynamic analysis of the signal.
Tone	This method is useful in disrupting UAVs' localization radars like SAR and provides precise control by applying it to a single or multiple tones.	To achieve good results, the placement of tones is crucial. However, this method is not very effective against frequency hopping spread spectrum (FHSS) systems.
Swept	To cover a broad range of frequencies, less power is needed. It is efficient against DSSS.	The performance of FHSS jamming can be unpredictable due to the changing jamming and signal tones, and as a result, mitigation strategies are being developed to counteract this.
Follower	It is efficient against FHSS.	Additional resources are required for analyzing the entire spectrum.
Smart	To jam FHSS and DSS signals, this method is the most reliable and power-efficient.	The technique requires prior knowledge of the target signal and analysis through technologies like SDR is necessary.

d. DoS attacks: The denial of service (DoS) attack is a frequently used type of attack against UAV networks. Figure 10.5 explains this attack, which is designed to slow down the information exchange by flooding the network with an excessive number of requests, which can overload and prevent legitimate requests from being serviced. This kind of attack can seriously impair the real-time communication of UAV networks, and even a minor DoS attack can have a detrimental effect. When a large-scale DoS attack or distributed denial of service (DDoS) attack occurs, the communication resources intended for the vehicle and infrastructure become saturated [17]. To execute a DoS attack, large packets are sent to the ground control station using a malicious source to disable the control signals. This results in the drone entering a lost link-state, which causes a malfunction in the data link. As a result, the drone operator cannot send or receive data signals from the UAV's flight control system, which

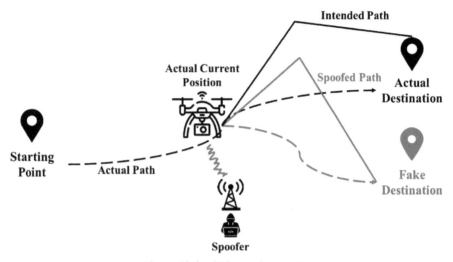

Figure 10.4 GPS spoofing attack.

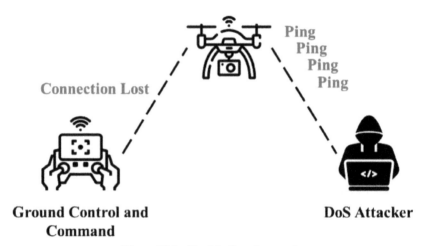

Figure 10.5 Denial of service attack.

disrupts the communication link and leads to a loss of control over the drone [18].

10.3 The Need for UAM

60% of world's population will be urban by 2030. As ground infrastructure becomes overcrowded, this growth in population will entail new transport options. A secure, sustainable, and convenient solution that exploits city airspace could aid such scenarios [19].

The objective of the UAM initiative is to develop a safe and effective air transportation system that operates at lower altitudes within urban and suburban areas and employs highly automated aircraft to transport passengers or goods. UAM is made possible by the use of VTOL technology. Low disk-loading rotors and non-traditional propulsion methods are used to reduce power and energy needs for short-range travel. With advancements in technology for structures, automation, energy, and design analysis, coupled with increasing population density and limited resources, this is an opportune time to explore new methods of transportation for people and goods.

a. Commercial inter-city (longer-range/thin haul):

Commercial inter-city or longer-range air mobility refers to the use of aerial vehicles to transport passengers between cities or regions. These vehicles may include helicopters, vertical takeoff and landing (VTOL) aircraft, or fixed-wing planes. Urban air mobility can help to reduce the time and cost associated with long-distance travel and provide an alternative to traditional ground transportation. The development of electric and hybrid-electric propulsion systems is making it possible to create quieter and more environmentally friendly vehicles for this purpose.

b. Cargo delivery:

UAM can be used for cargo delivery in urban areas, particularly for time-sensitive and high-value goods. Delivery drones and vertical takeoff and landing (VTOL) aircraft can provide quick and efficient transportation of goods within cities, reducing road congestion and improving delivery times. Companies are exploring the use of UAM for last-mile delivery, where goods are transported from a central hub to a final destination.

c. Public services:

UAM can be used to provide public services such as emergency medical transportation, law enforcement surveillance, and disaster relief. Aerial vehicles can reach remote or difficult-to-access areas quickly and efficiently, providing vital services when ground transportation is not possible. For example, drones equipped with medical supplies or equipment can transport them to accident or disaster sites, where they can be used to provide first aid or other medical assistance.

d. Private/recreational vehicles:

UAM can also be used for private and recreational purposes, such as sightseeing tours or personal transportation. Private air taxis and VTOL aircraft could offer a faster and more convenient alternative to ground transportation in urban areas. However, regulatory and safety challenges will need to be addressed before widespread adoption of private urban air mobility vehicles can occur.

The previous four points describe general applications of UAM [20], while the following points focus specifically on the advantages of UAM in emergency response [21].

e. Faster response time:

UAM can improve emergency response by allowing aircraft specifically designed for this purpose to bypass traffic congestion and reach the scene of emergencies much quicker. A study by Airbus found that the use of eVTOL aircraft could result in response times that are up to 50 times faster than traditional ground-based transportation. Flying over traffic instead of being slowed down by it can be particularly beneficial when transporting patients in need of urgent care.

f. Increased efficiency:

UAM aircraft can increase the efficiency of emergency response by flying above obstacles like buildings, power lines, and highways. During natural disasters, where ground transportation is limited, UAM can provide a much faster and safer means for emergency responders to reach affected areas and provide aid.

g. Improved safety:

UAM aircrafts are safer than other forms of transportation because of their advanced safety features like onboard parachutes, sensors, and technology to assist pilots. Traditional emergency responders often face danger while traveling in large trucks or ambulances on the road, making UAMs a safer option for them.

h. Easier to reach remote areas:

Air ambulances or flying cars can be used to reach previously inaccessible locations, such as bridges, tall buildings, and remote areas. For example, if there is a car accident on a bridge, emergency responders can still reach the scene using drones as air ambulances, even in bad weather. Air ambulances can also reach remote areas where traditional transportation is limited. This can improve emergency services' ability to respond to accidents and disasters in these areas.

i. Organs transportation:

A significant advantage of using air mobility vehicles for emergency services is the fast transportation of transplant organs from the donor hospital to the recipient hospital. The longer an organ is outside the body, the less likely it is to survive the transplant. With UAM service, organs can be transported in minutes rather than the hours it would take with a traditional ambulance, increasing the chances of a successful transplant.

j. Flying ambulance:
Flying ambulances can transport patients quickly across a city, bypassing traffic congestion and providing faster access to emergency medical care. This is especially beneficial in densely populated areas where traditional ground-based ambulances can be slowed down. Flying ambulances are also more efficient and compact than traditional ambulances, requiring fewer vehicles to transport patients and medical personnel. Additionally, they can deliver medical supplies such as vaccines and antidotes rapidly, reducing the spread of diseases during supply chain disruptions or natural disasters. UAM provides a range of advantages for emergency services, improving the speed and efficiency of personnel and equipment transportation.

10.4 Major Challenges in Implementing UAM

10.4.1 Autonomy

The use of automation and autonomy is crucial for the success of air mobility solutions and will require significant investments in validation and optimization of systems and vehicles, as well as testing infrastructure such as wind tunnels, hybrid and electric propulsion, and high-performance computing. These technologies also have connections to artificial intelligence and digitalization; so collaboration between fields is necessary [22].

10.4.2 Public acceptance and environment

The examination of the social and economic impact is vital for the future of urban mobility. Furthermore, it is primarily the issues of noise and weather that must be addressed to gain public support and for the design of new vehicles and systems.

10.4.3 Unmanned aircraft system traffic management (UTM)

The integration of UAM systems, including the field of UTM/U-space, will be essential for the future and economic success of unmanned and autonomous aircraft systems. This will involve a focus on system certification, regulations, societal acceptance, and testing. The management of air traffic is a complex system that involves both human and machine. It is the responsibility of air traffic control (ATC) to regulate and supervise aircraft in the airspace. The system employs technologies such as navigation systems, speech synthesis, sensors, actuators, and intelligent systems to increase automation in the region and minimize delays, pollution, noise, and accidents [23]. The ATC also manages air traffic congestion. The primary objective of ATC is to ensure the safety of aircraft, pilots, passengers, and flight attendants. This is achieved by enabling communication between all parties involved throughout the entire flight.

10.4.4 The unavailability of landing sites (vertiports)

Currently, many cities have air traffic control systems that monitor helicopter activity; these systems can be adapted to track the movements of numerous eVTOLs (electric vertical takeoff and landing) aircrafts as well. As eVTOL becomes a more popular mode of transportation, with thousands of flying taxis in operation, a new airspace management system will be necessary. This system should be able to detect and precisely locate nearby eVTOLs. The implementation of mass transportation systems will also require cities to plan for public vertiports and offer discounted prices for users of the new transportation system [24].

10.4.5 Safety concerns

The varied use of urban airspace necessitates the adoption of innovative security measures to ensure the safe coordination of eVTOLs in the air. This includes implementing a high-speed, low-latency network connection to ensure seamless transmission of mission-critical information related to an eVTOL's flight trajectory. Other advanced technologies, such as artificial intelligence, may also be necessary to manage flight plans and detect conflicts. Additionally, the large amounts of data being transferred raise concerns about the collaboration of different parties involved in flight management. To ensure safe and efficient UAM within cities and between cities, full compliance with safety regulations is essential. Lastly, UAM manufacturers must use robust and long-lasting materials and components to ensure safe operation in harsh weather conditions. Only by addressing these three security aspects (airspace, data, and manufacturing) can passenger safety be guaranteed.

10.4.6 Ground infrastructure

The digitization of UAM heavily relies on ground infrastructure, which faces several challenges. It requires the adoption of physical assets to enable full automation. Designing the infrastructure requires a thorough analysis of the specific city to ensure its effectiveness.

10.4.7 Noise

Reducing noise is a significant concern for aviation, and there are several methods for achieving it. These include modifications to engines, flight paths, land usage, and urban development. Noise can be caused by both acoustic and non-acoustic factors, and addressing them is crucial to mitigate noise pollution.

10.5 Safety and Security Challenges

The previous section discussed the challenges of implementing UAM in real time, while this section focuses specifically on one of those challenges: safety and security [25].

10.5.1 Safe autonomy

The use of autonomy in drones and air taxis could open up new possibilities for increasing their numbers, making the services more widely accessible. Additionally, from a safety standpoint, if tasks typically performed by a pilot, such as emergency management, can be automated, it could increase the overall safety of the system. One crucial task of a pilot is to oversee the flight and take action in case of a failure. Without a pilot on board, there is no fallback option in case of onboard system failure. It is therefore essential to automate this fallback and monitor the overall system state. If the autonomous function can determine the best course of action and execute it in case of an emergency, it would greatly increase safety. This type of autonomous function could also be integrated into manned flight to further enhance safety.

10.5.2 Reliable multi-sensor navigation

Urban air mobility demands dependable navigation information to safely operate a high number of vehicles in shared airspace. It is anticipated that UAM navigation systems will rely on redundant sensors and data in order to meet the necessary levels of accuracy, integrity, availability, and continuity. This is challenging due to several factors, such as the limitations of size, power, and cost that may prevent UAM vehicles or cargo drones from utilizing high-grade sensors. Additionally, for various UAM applications and types of operations, such as takeoff or landing, precise guidance must be ensured even in challenging environmental conditions like poor GNSS visibility. New technologies that lack standardization and a history of in-flight use must also be taken into account. To ensure safe and smooth service operation, the navigation system must provide sufficient accuracy and at the same time, high reliability or integrity of the positioning result.

10.5.3 Robust and efficient communication

Robust and efficient communication is critical for the safe and effective operation of unmanned aerial vehicles (UAVs), especially in the context of urban air mobility (UAM). However, there are several challenges that must be addressed to achieve reliable communication in these environments. One significant challenge is the lack of an existing communication system designed specifically for UAM. Additionally, the high mobility of vehicles in the 3D plane, non-line-of-sight (NLOS) conditions, and strong multipath propagation make it challenging to establish reliable and consistent communication links. Moreover, efficient usage of shared resources is crucial for effective communication in UAM.

To address these challenges, several research gaps need to be addressed. These include the requirements for collision avoidance, latency, bandwidth, and availability. Additionally, research is needed to develop communication concepts and a channel model for UAM scenarios that consider both air-to-air (A2A) and air-to-infrastructure (A2I) communication. Finally, the propagation characteristics in urban environments must be well understood to ensure reliable and efficient communication.

10.5.4 U-space and safe air traffic

U-space and safe air traffic are critical considerations for the safe and efficient operation of unmanned aerial vehicles (UAVs) in the low-altitude airspace. However, there are several challenges that must be addressed to ensure the success of these efforts. One significant challenge is the lack of availability of U-space services, which have experienced delays in their rollout. Additionally, the low-altitude airspace is frequented by both birds and drones, which poses a safety risk to UAVs and requires specialized on-board sensors to detect and identify them.

To address these challenges, several research gaps must be addressed. These include the development of the required U-space services, such as vertidrome management, collision avoidance, and flight path prediction of birds and drones. Additionally, research is needed to identify the on-board sensor requirements for detecting and identifying birds and drones and evaluate the impact resistance required to avoid damage in case of a collision. Flight demonstrations can also help identify the required information from U-space and provide insights into the efficacy of these services.

10.5.5 Cyber−physical safety and security

Cyber−physical safety and security is a critical consideration for the safe and efficient operation of unmanned aerial vehicles (UAVs). However, there are several challenges that must be addressed to ensure the success of these efforts. One significant challenge is the prevention, detection, response, and mitigation of diverse attack vectors, including both physical and cyber threats. Additionally, shared situational awareness is crucial for efficient crisis resolution.

To address these challenges, several research gaps must be addressed. These include the need for an adapted definition of "aviation security" that considers the unique challenges posed by UAVs and the switch from attackers being transport system users to attackers acting remotely from anywhere. Research is also needed to analyze the overall system and develop critical operation procedures for takeoff and landing. Moreover, physical and cyber-attacks and their combinations must be considered to develop effective strategies for prevention, detection, response, and mitigation.

10.6 Abusive Use of Drones

Unintentional abuse of a drone refers to actions taken by the drone pilot that are based on ignorance and do not consider case law. This can result in harm to individuals, property, finances, or ideology. On the other hand, intentional misuse involves the drone pilot (and possibly their principal) purposely engaging in damaging or financially beneficial actions. Those who engage in intentional misuse include individuals with personal motives, organized criminals, and terrorist networks. This type of misuse can also result in harm to individuals, property, finances, or ideology [26].

10.6.1 Classes of abuse

10.6.1.1 Attack

One potential threat posed by drones is an attack by a swarm of unmanned aircraft system (UAS) that could be used for collision or crashing into targets. Such an attack could cause significant damage to structures or other physical targets, and potentially even result in loss of life. In addition to physical attacks, drones could also be used to carry and deploy toxins or explosives. This could be particularly dangerous if a large number of drones are used simultaneously, as it would be difficult for authorities to track and intercept all of them before they reach their targets. Figure 10.6 clearly explains the classes of abuse.

To prevent these types of attacks, it is essential to develop effective countermeasures, such as advanced surveillance systems, anti-drone technology, and emergency response plans. Additionally, regulations and laws should be established to prevent the misuse of drones for criminal purposes, and penalties for violating these regulations should be severe enough to act as a deterrent. Ongoing research and development of counter-drone technology and strategies will also be critical in ensuring the safety and security of individuals and property in the face of potential drone attacks.

10.6.1.2 Espionage

Drones can also pose a threat in terms of espionage, as they can be used to gather sensitive information through the use of advanced sensing systems.

Figure 10.6 Classes of abuse.

These sensing systems can include high-resolution cameras, thermal imaging technology, and other sensors that can capture a range of data from a distance.

Such data collection can be used for a variety of purposes, including surveillance of individuals or groups, gathering of sensitive information such as trade secrets or classified documents, or even theft of personal data. This raises concerns about privacy violations, as the use of drones in this way can potentially capture sensitive personal information.

To prevent such espionage, it is important to establish regulations around the use of drones for data collection purposes and to ensure that such use is subject to strict privacy protections. This could include requiring drones to be registered, establishing no-fly zones around sensitive areas, and mandating that drone operators obtain special permits or licenses for certain types of drone use.

Another important strategy is to develop counter-drone technology that can detect and disrupt the operation of drones used for espionage purposes. This could include using jamming devices to block drone communications or even developing drones that can intercept and disable other drones in flight.

Ultimately, it will be critical to balance the potential benefits of drone technology with the need to protect individual privacy and security in order to prevent the misuse of drones for espionage purposes.

10.6.1.3 Sabotage

Drones can also be used for sabotage by carrying a "dirty" load that can damage or render an item unserviceable. This can include the delivery of corrosive or harmful materials, or even the use of explosives to cause physical damage to a target.

The use of drones for sabotage raises significant safety concerns, as it can potentially cause harm to individuals and damage to property. To prevent such sabotage, it is important to establish regulations around the use of drones for delivery purposes and to ensure that such use is subject to strict safety protocols.

This could include requiring drones to be registered, establishing no-fly zones around sensitive areas, and mandating that drone operators obtain special permits or licenses for certain types of drone use. Additionally, security measures such as screening and monitoring of cargo can be implemented to detect and prevent the transportation of harmful materials.

Overall, it is important to be vigilant in identifying potential threats posed by the use of drones for sabotage, and to take proactive steps to mitigate these risks in order to ensure the safe and responsible use of drone technology.

10.6.1.4 Intrusion

Drones can be used for intrusion, particularly for smuggling prohibited items. Smuggling can include the transportation of illegal drugs, weapons, and other contraband, as well as the smuggling of people across borders.

The use of drones for smuggling poses significant challenges for law enforcement agencies, as they are difficult to detect and can fly undetected in remote or hard-to-reach areas. To combat the use of drones for smuggling, authorities may employ a range of strategies, such as establishing no-fly zones, deploying surveillance equipment to detect and track drones, and developing new technologies to intercept or disable drones.

In addition to these security measures, there is also a need to address the root causes of smuggling, such as poverty, corruption, and political instability, which can fuel demand for illegal goods and services. This may require international cooperation and coordinated efforts between law enforcement agencies to disrupt smuggling networks and reduce demand for illegal goods.

Overall, the use of drones for smuggling highlights the need for ongoing efforts to develop effective security and law enforcement strategies to combat the misuse of drone technology.

10.7 Air Taxi Related Threats

This section discusses the potential safety and security threats related to the use of air taxis, including issues such as cyber-attacks, physical attacks, and accidents.

10.7.1 Shooting

The threat of shooting against air taxis mainly involves small sports or hunting arms. This could be due to a variety of reasons, such as criminal activity, terrorism, or even accidental discharge. Shooting at air taxis could cause significant damage and potentially result in a catastrophic event, putting passengers and bystanders at risk. To address this threat, air taxi operators may implement security measures such as increased surveillance, perimeter protection, and communication with local law enforcement agencies. Additionally, advances in drone detection and countermeasures technology could provide an effective means of identifying and neutralizing potential threats.

10.7.2 Dirty load

Air taxi services could be targeted with a "dirty load" attack, which involves the transport of explosives, toxic materials, drugs, or other hazardous substances. This could pose a significant risk to passengers, crew, and the general public, potentially resulting in loss of life or widespread damage. To mitigate this threat, air taxi operators may implement strict security protocols and screening measures to prevent prohibited items from being brought on board. Additionally, advanced sensing and detection technologies could be employed to identify and intercept potential threats before they can cause harm. Ongoing collaboration and information sharing between air taxi operators and law enforcement agencies is also critical to prevent and respond to dirty load attacks.

10.7.3 Jamming

Jamming is a threat to air taxi services that involves the deliberate interference with the wireless communication link between the unmanned aircraft and its ground control station. This can disrupt or disable the downlink, leading to a loss of situational awareness and control over the vehicle. This could result in the air taxi becoming uncontrollable and potentially causing a crash. To address this threat, air taxi operators may employ anti-jamming technology and backup communication systems to maintain control in the event of an attack. Furthermore, collaboration with government agencies to locate and neutralize sources of jamming could also be effective in preventing these attacks from occurring.

10.7.4 Spoofing

Spoofing is a threat that involves the manipulation of GPS signals to deceive the unmanned aircraft's navigation system. This can cause the vehicle to deviate from its intended course or to fly toward a new destination. The attackers can take control of the air taxi and use it for malicious purposes such as espionage, sabotage, or terrorist attacks. To prevent spoofing attacks, air taxi operators can employ GPS receivers that are resistant to spoofing, monitor the GPS signal for anomalies, and have backup navigation systems that do not rely on GPS. Additionally, government regulations and standards can be developed to require the use of more secure GPS technologies.

10.7.5 Overheating

Overheating is a potential threat that can result in the thermal runaway of battery cells, leading to a fire or explosion. This can be caused by a variety of factors such as mechanical damage, electrical shorts, or exposure to high temperatures. Electromagnetic pulse (EMP) weapons are also a concern, as they can cause electronic components to malfunction and overheat. To mitigate the risk of overheating, air taxi operators can use batteries with built-in thermal management systems, monitor battery temperature during operation, and limit the number of charging cycles. In addition, shielding against EMP can be employed in air taxi systems to prevent electromagnetic interference.

10.7.6 Kamikaze

Kamikaze attacks to the use of remotely piloted or programmed drones to target and collide with air taxis in flight. This type of attack can be carried out by individuals with malicious intent, such as terrorists or criminals. To mitigate this threat, air taxi operators can employ technologies such as collision avoidance systems and radar to detect and avoid incoming threats. Additionally, air traffic control and law enforcement agencies can monitor airspace for suspicious activity and take appropriate action to intercept and neutralize any threats. Regular training and simulation exercises can also help prepare air taxi operators and first responders to respond to potential kamikaze attacks.

10.8 Security Measures

This section describes the different types of security measures that can be used to protect UAVs and the data they carry [27].

- **Passive, preventive:** This type of security measure involves using structural or privacy measures to prevent unauthorized access to UAVs or their data. Structural measures may include physical barriers or locks to protect the UAV or its storage devices, while privacy measures may include encryption or other methods to protect the data.
- **Active, disturbing:** This type of security measure involves using active methods to disrupt or disturb the UAV's signals or communications. Examples include jamming the signal to disrupt communication between the UAV and its controller, spoofing the signal to take control of the UAV, hijacking the UAV, or intercepting the UAV's data.

- **Active, preventive:** This type of security measure involves using active methods to prevent attacks or unauthorized access. Access control measures may be used to limit who has access to the UAV or its data. Modulating the signal strength or data transfer rate can also make it more difficult for attackers to intercept or hijack the UAV.
- **Active, destroying:** This type of security measure involves using active methods to physically destroy or disable the UAV. Examples include using electromagnetic pulse weapons or laser guns to destroy the UAV or using glue guns or high-pressure water cannons to disable the UAV's propellers or other components.

10.8.1 Safety and security

Safety and security are critical aspects of urban air mobility (UAM) when operating over populated areas. The following is an elaboration of each of the headings mentioned.

1. **Adverse weather and airflow conditions at low altitudes:** UAM operations are subject to various weather conditions and airflow patterns, which can impact the safety and security of operations. Adverse weather conditions such as strong winds, heavy rain, and thunderstorms can affect the performance and stability of UAM vehicles, and lead to accidents or crashes. Airflow conditions at low altitudes, such as turbulence or downdrafts, can also impact the safety of UAM operations.
2. **Human factors and automation, collision, and avoidance:** Human factors play a critical role in the safety and security of UAM operations. It is important to consider how pilots, operators, and ground crew interact with the technology and systems used in UAM operations. Automation can be used to reduce the workload on operators and pilots, but it also introduces new challenges such as automation bias and complacency. Collision and avoidance is another important aspect of UAM safety and security, as UAM vehicles share the same airspace as manned aircraft and other UAM vehicles.
3. **Electromagnetic compatibility:** UAM vehicles rely on various communication and navigation systems, which can be disrupted by electromagnetic interference from other devices or sources. It is important to ensure that UAM systems are designed to operate in an electromagnetically compatible environment, to minimize the risk of interference and ensure safe and secure operations.

4. **Detection and surveillance of physical and cyber threats, prevention, preparedness, response and recovery from threats, including intentional interference and misuse of urban air mobility:** The security of UAM operations is also critical, as they are subject to various physical and cyber threats. It is important to detect and monitor potential threats, and implement measures to prevent or mitigate them. This may include surveillance systems, access control measures, encryption, and other security protocols. It is also important to be prepared to respond to and recover from security incidents, including intentional interference or misuse of UAM vehicles.

5. **Other relevant hazards and threats in an operation-centric and risk-based approach:** UAM operations are subject to a range of other hazards and threats, which may include bird strikes, power lines, and other obstacles. A risk-based approach is important for identifying and assessing these hazards, and implementing appropriate mitigation measures. This may include route planning, obstacle detection and avoidance systems, and other safety measures.

10.9 Summary

Urban air mobility (UAM) is an emerging field that has the potential to revolutionize urban transportation by providing a faster, more efficient, and sustainable mode of travel. While the concept of UAM has been around for decades, recent technological advancements in electric propulsion, autonomy, and materials science have made it possible to develop practical UAM solutions. Despite the numerous benefits that UAM promises to offer, there are still several challenges that need to be addressed, including safety, security, regulation, infrastructure, and public acceptance. However, with the increasing interest and investment in UAM, it is expected that these challenges will be overcome, and we will see the widespread adoption of UAM services in the near future, transforming the way we travel in urban areas. This chapter has presented a comprehensive overview of UAM, its need, and the challenges that come along with it. The chapter has discussed the safety and security measures required for the successful implementation of UAM, including various physical and cyber threats that can compromise the system's integrity. The chapter has also touched upon the potential abusive use of drones and air taxis, and the measures that can be taken to prevent such situations. Overall, the chapter highlights the importance of safety and security in the UAM industry and the need for comprehensive measures to address these challenges. It is clear that as the UAM industry continues to grow, safety and security will remain a critical factor, and continued research and development in this area will be necessary to ensure the success of UAM in the future.

References

[1] Cokorilo, O. (2020). Urban air mobility: safety challenges. Transportation research procedia, 45, 21-29.

[2] https://www.statista.com/statistics/564717/airline-industry-passenger-traffic-globally/

[3] https://bisresearch.com/industry-report/global-urban-air-mobility-market.html

[4] Cohen, A., Guan, J., Beamer, M., Dittoe, R., & Mokhtarimousavi, S. (2020). Reimagining the future of transportation with personal flight: Preparing and planning for urban air mobility.

[5] Hill, B. P., DeCarme, D., Metcalfe, M., Griffin, C., Wiggins, S., Metts, C., ... & Mendonca, N. L. (2020). Uam vision concept of operations (conops) uam maturity level (uml) 4.

[6] Nagori, C. (2020). Unmanned Aerial Manipulators in Construction-Opportunities and Challenges (Doctoral dissertation, Virginia Tech).

[7] Nordin, M. H., Sharma, S., Khan, A., Gianni, M., Rajendran, S., & Sutton, R. (2022). Collaborative unmanned vehicles for inspection, maintenance, and repairs of offshore wind turbines. Drones, 6(6), 137.

[8] Shafiee, M., Zhou, Z., Mei, L., Dinmohammadi, F., Karama, J., & Flynn, D. (2021). Unmanned aerial drones for inspection of offshore wind turbines: A mission-critical failure analysis. Robotics, 10(1), 26.

[9] Smith, K. W. (2015). Drone technology: Benefits, risks, and legal considerations. Seattle J. Envtl. L., 5, 291.

[10] Kong, P. Y. (2021). A survey of cyberattack countermeasures for unmanned aerial vehicles. IEEE Access, 9, 148244-148263.

[11] Li, K., Kanhere, S. S., Ni, W., Tovar, E., & Guizani, M. (2019, June). Proactive eavesdropping via jamming for trajectory tracking of UAVs. In 2019 15th International Wireless Communications & Mobile Computing Conference (IWCMC) (pp. 477-482). IEEE.

[12] Chamola, V., Kotesh, P., Agarwal, A., Gupta, N., & Guizani, M. (2021). A comprehensive review of unmanned aerial vehicle attacks and neutralization techniques. Ad hoc networks, 111, 102324.

[13] Šimon, O., & Götthans, T. (2022). A Survey on the Use of Deep Learning Techniques for UAV Jamming and Deception. Electronics, 11(19), 3025.

[14] Benkraouda, H., Barka, E., & Shuaib, K. (2018). Cyber-attacks on the data communication of drones monitoring critical infrastructure. Comput. Sci. Inf. Technol, 8(17), 83-93.

[15] Jacobsen, R. H., & Marandi, A. (2021, November). Security threats analysis of the unmanned aerial vehicle system. In MILCOM 2021-2021 IEEE Military Communications Conference (MILCOM) (pp. 316-322). IEEE.

[16] Wei, X., Sun, C., Lyu, M., Song, Q., & Li, Y. (2022). ConstDet: Control Semantics-Based Detection for GPS Spoofing Attacks on UAVs. Remote Sensing, 14(21), 5587.

[17] Mairaj, A., & Javaid, A. Y. (2022). Game theoretic solution for an Unmanned Aerial Vehicle network host under DDoS attack. Computer Networks, 211, 108962.

[18] Mekdad, Y., Aris, A., Babun, L., El Fergougui, A., Conti, M., Lazzeretti, R., & Uluagac, A. S. (2023). A survey on security and privacy issues of UAVs. Computer Networks, 224, 109626.

[19] https://www.airbus.com/en/innovation/zero-emission-journey/urban-air-mobility

[20] https://www.faa.gov/uas/advanced_operations/urban_air_mobility

[21] https://businessingmag.com/17420/equipping/air-mobility/#:~:text=Urban%20air%20mobility%20can%20help%20emergency%20responders%20reach,and%20quickly%20reach%20the%20scene%20of%20the%20emergency

[22] https://futuresky.eu/themes/urban-air-mobility/

[23] McShane, W. R., & Roess, R. P. (1990). Traffic engineering.

[24] https://blog.crouzet.com/4-key-challenges-in-implementing-urban-air-mobility/

[25] Torens, C., Volkert, A., Becker, D., Gerbeth, D., Schalk, L., Garcia Crespillo, O., ... & Dauer, J. (2021). HorizonUAM: Safety and security considerations for urban air mobility. In AIAA Aviation 2021 Forum (p. 3199).

[26] Holger Zeiser, "Security Aspects of Urban Air Mobility: Are We Prepared?", European Aviation Security Centre, Civitas Forum, 2019.

[27] Adam, C., & Susan, S. (2021). Urban air mobility: Opportunities and obstacles. Transportation Sustainability Research Centre, 702-709

Index

About the Editors

Vishnu Kumar Kaliappan is working as Professor in Computer Science and Engineering Department at KPR Institute of Engineering and Technology & Konkuk University Seoul, South Korea and has 15.8 years of Teaching and Research Experience. He received his Ph.D in Computer and Information Communication Engineering from Konkuk University, Seoul, South Korea during 2012 and received M.Tech in Communication Engineering from VIT University, Vellore, India. He is an Editorial Manager at ISIUS (International Society of intelligent Unmanned System), Korea. He worked the project under KARI (Korean Aerospace Research Institute), Degu, South Korea and at CABS (Centre for Air Born System), DRDO, Bangalore, India. He received one of the Korean prestigious Scholarship IITA (International Information Technology Admission) from Ministry of Information Technology, Seoul, South Korea in the 2007-2012. He has published more than 70 peer reviewed Journals, conference and book chapters. The focus of his research is on Reinforcement Learning, bio mimetic algorithms, Cyber Physical System, Hardware in the Loop Simulation (HILS) and Control algorithms for unmanned aerial vehicles. He has been acted as reviewer and editorial member for more 70 international conference and journals.

Mohana Sundaram Kuppusamy received B.E. degree in Electrical and Electronics Engineering from University of Madras in 2000, M.Tech degree in High Voltage Engineering from SASTRA University in 2002 and Ph.D. degree from Anna University, India in 2014. His research interests include Intelligent controllers, Power systems, Embedded system and Power electronics. He has completed funded project of worth Rs.30 .79 lakhs sponsored by DST, Government of India. Currently he is working as a Professor in EEE department at KPR Institute of Engineering and Technology, India. He has produced 04 Ph.D candidates under his supervision in Anna University, Chennai. He has published three books and serving as reviewer for IEEE, Springer and Elsevier journals. He is an active member of IE, ISTE and IAENG. He has published around 47 papers in International journals.

Dugki Min received the B.S. degree in industrial engineering from Korea University, Seoul, South Korea, in 1986, and the M.S. and Ph.D. degrees in computer science from Michigan State University, East Lansing, MI, USA, 1991 and 1995, respectively. He is a Professor with the Department of Computer Science and Engineering, College of Engineering, Konkuk University, Seoul, where he is the Head of Distributed Multimedia Systems Laboratory. His current research interests include multiagent systems, deep learning and deep reinforcement learning, intelligent Internet of Things, cyber-physical systems, digital twin systems simulation, intelligent big data analysis, intelligent fog computing, proactive provisioning for cloud computing, distributed and parallel computing, software architecture design, and performance, and dependability analysis.

For Product Safety Concerns and Information please contact our EU
representative GPSR@taylorandfrancis.com
Taylor & Francis Verlag GmbH, Kaufingerstraße 24, 80331 München, Germany

www.ingramcontent.com/pod-product-compliance
Ingram Content Group UK Ltd.
Pitfield, Milton Keynes, MK11 3LW, UK
UKHW021821240425
457818UK00001B/16

* 9 788770 226783 *